Revisiting Value Co-creation and Co-destruction in Tourism

This book assists the better understanding of value co-creation and co-destruction in tourism development by bringing together different perspectives and disciplines. It provides some examples of how value can be co-created or co-destroyed within the context of tourism.

Tourism is susceptible to uncertainty and incidents that can directly impact the supply and demand of its discretionary products and services. Consensus has been reached among practitioners and academics that consumer experience is more important than ever for enterprises as well as destinations, as the sector has become globalized, reached maturity and become highly competitive. Still, the pathway to success (or failure) lies within the overall satisfaction of visitors and tourists, which heavily depends on perceived value; a concept that can be co-created or co-destroyed by the very interaction between all social actors and stakeholders involved. Value creation or destruction is critical not just for traditional supply and demand, but also for an array of actors across value and distribution chains (including, for example, staff and intermediaries across the networks).

The book will be of great value to scholars, students and policymakers interested in tourism studies and practices and service management as well as professionals in the field of tourism management.

The chapters were originally published as a special issue of the journal, *Tourism Planning & Development*.

Elina (Eleni) Michopoulou is Associate Professor in Business Management at University of Derby, UK. Eleni has published over 60 academic journal articles, book chapters and conference papers, sits on the editorial board of over 10 high impact academic journals and is the editor-in-chief of *International Journal of Spa and Wellness*.

Nikolaos Pappas is Associate Professor in Tourism, Hospitality and Events and the Director of CERTE (Centre for Research in Tourism Excellence) at the University of Sunderland, UK. His research interests focus on tourism planning and development, and risk and crisis management.

Elena Cavagnaro is Professor of Sustainability in Hospitality and Tourism at NHL Stenden University of Applied Sciences. Elena has consulted several organizations in sectors such as hospitality, retail and health care on sustainability strategy and implementation. Her research focuses on sustainability within social, organizational and individual spheres.

Revisiting Value Co-creation and Co-destruction in Tourism

Edited by
Elina (Eleni) Michopoulou, Nikolaos Pappas and Elena Cavagnaro

Routledge
Taylor & Francis Group
LONDON AND NEW YORK

First published 2022
by Routledge
2 Park Square, Milton Park, Abingdon, Oxon OX14 4RN

and by Routledge
605 Third Avenue, New York, NY 10158

Routledge is an imprint of the Taylor & Francis Group, an informa business

British Library Cataloguing in Publication Data
A catalogue record for this book is available from the British Library

ISBN13: 978-1-032-15750-4 (hbk)
ISBN13: 978-1-032-15751-1 (pbk)
ISBN13: 978-1-003-24553-7 (ebk)

DOI: 10.4324/9781003245537

Typeset in Myriad Pro
by Newgen Publishing UK

Publisher's Note
The publisher accepts responsibility for any inconsistencies that may have arisen during the conversion of this book from journal articles to book chapters, namely the inclusion of journal terminology.

Disclaimer
Every effort has been made to contact copyright holders for their permission to reprint material in this book. The publishers would be grateful to hear from any copyright holder who is not here acknowledged and will undertake to rectify any errors or omissions in future editions of this book.

Contents

Citation Information

The chapters in this book were originally published in the *Tourism Planning & Development*, volume 18, issue 2 (2021). When citing this material, please use the original page numbering for each article, as follows:

Introduction

Revisiting Value Co-creation and Co-destruction in Tourism
Elena Cavagnaro, Eleni Michopoulou and Nikolaos Pappas
Tourism Planning & Development, volume 18, issue 2 (2021), pp. 121–124

Chapter 1

Revisiting Online Tourism Forums as Vehicles for Value Co-destruction
Aaron Tham and Mingzhong Wang
Tourism Planning & Development, volume 18, issue 2 (2021), pp. 125–146

Chapter 2

The Role of Industry 4.0 Tools on Museum Attributes Identification: An Exploratory Study of Thyssen-Bornemisza National Museum (Madrid, Spain)
Alicia Orea-Giner, Carmen De-Pablos-Heredero and Trinidad Vacas-Guerrero
Tourism Planning & Development, volume 18, issue 2 (2021), pp. 147–165

Chapter 3

Generating and Sustaining Value Through Guided Tour Experiences' Co-Creation at Heritage Visitor Attractions
Kamila Bezova and Iride Azara
Tourism Planning & Development, volume 18, issue 2 (2021), pp. 226–244

Chapter 4

Tourist–Tourist Social Interaction in the Co-creation and Co-destruction of Tourism Experiences among Chinese Outbound Tourists
Xing Han, Carolus L. C. Praet and Liyong Wang
Tourism Planning & Development, volume 18, issue 2 (2021), pp. 189–209

Chapter 5

Value Co-Creation and Co-Destruction: Considerations of Spa Servicescapes
Louise Buxton and Eleni Michopoulou
Tourism Planning & Development, volume 18, issue 2 (2021), pp. 210–225

Chapter 6

Destination Image Co-creation in Times of Sustained Crisis
Kyriaki Glyptou
Tourism Planning & Development, volume 18, issue 2 (2021), pp. 166–188

Chapter 7

Co-creating Value in Desert Tourism Experiences
Eleni Michopoulou, Idrees Al-Qasmi and Claudia Melpignano
Tourism Planning & Development, volume 18, issue 2 (2021), pp. 245–265

Notes on Contributors

Idrees Al-Qasmi, Centre for Contemporary Hospitality and Tourism, University of Derby, Derbyshire, UK.

Iride Azara, Centre for Contemporary Hospitality and Tourism, University of Derby, Derby, UK.

Kamila Bezova, Centre for Contemporary Hospitality and Tourism, University of Derby, Derby, UK.

Louise Buxton, Centre for Contemporary Hospitality and Tourism, College of Business Law and Social Science, University of Derby, UK.

Elena Cavagnaro, NHL Stenden University, Leeuwarden, Netherlands.

Carmen De-Pablos-Heredero, Social Sciences Faculty, Universidad Rey Juan Carlos, Madrid, Spain.

Kyriaki Glyptou, School of Events, Tourism and Hospitality Management, Leeds Beckett University, Leeds, UK.

Xing Han, Graduate School of Modern Commerce, Otaru University of Commerce, Hokkaido, Japan.

Claudia Melpignano, Centre for Contemporary Hospitality and Tourism, University of Derby, Derbyshire, UK.

Eleni Michopoulou, Centre for Contemporary Hospitality and Tourism, College of Business Law and Social Science, University of Derby, UK.

Alicia Orea-Giner, Social Sciences Faculty, Universidad Rey Juan Carlos, Madrid, Spain; EIREST, Université Paris 1 Panthéon- Sorbonne, Paris, France.

Nikolaos Pappas, University of Sunderland, Sunderland, UK.

Carolus L. C. Praet, Department of Commerce, Otaru University of Commerce, Hokkaido, Japan.

Aaron Tham, USC Business School, University of the Sunshine Coast, Sunshine Coast, Australia.

Trinidad Vacas-Guerrero, Social Sciences Faculty, Universidad Rey Juan Carlos, Madrid, Spain.

Liyong Wang, Department of Commerce, Otaru University of Commerce, Hokkaido, Japan.

Mingzhong Wang, USC Business School, University of the Sunshine Coast, Sunshine Coast, Australia.

Introduction

Revisiting Value Co-creation and Co-destruction in Tourism

As COVID-19 has shown in a way unimaginable before it hit, tourism is susceptible to uncertainty and incidents that can directly impact the supply and demand of its discretionary products and services. Before the pandemic, consensus had been reached among practitioners and academics that consumer experience is more important than ever for enterprises as well as destinations, as the sector had become globalized, reached maturity and became highly competitive. Tourism came to a grinding halt due to the pandemic and recovery may take years. Still, the pathway to success (or failure) lies on the overall satisfaction of visitors and tourists, which heavily depends on perceived value; a concept that can be co-created or co-destroyed by the very interaction between all social actors and stakeholders involved. Value creation or destruction is critical not just for traditional supply of and demand for, but also for an array of actors across value and distribution chains (including for example staff and intermediaries across the networks). The special issue's aim was to assist the better understanding of value co-creation and co-destruction in tourism development by bringing together different perspectives and disciplines.

Judging from the diversity of the theoretical perspectives of the articles collected in this issue and the richness of the presented findings the special issue has indeed achieved its aim. Yet some real trends could be distinguished: the relevance of online communication and information; the importance of interpersonal encounters and social interaction for value co-creation and co-destruction in tourism; and the challenges in the design and delivery process of co-created experiences.

Information, communication and automation technology in a people-oriented sector

The importance of user generated content on social media for value co-creation and, particularly, co-destruction in tourism is underlined by Aaron Tham in his study entitled 'Revisiting Online Tourism Forums as Vehicles for Value Co-Destruction'. Tham explores trolling episodes on Lonely Planet Thorn Tree forums and the response of troll's victims. Trolling episodes bait, shame, and antagonize other social media users by framing seemingly innocent content in a provoking way or insolently responding to other users' content. Trolling is a form of cyberbullying and has got the attention of, for example, education but not yet of tourism researchers. However, trolling episodes potentially inflict serious damage to a destination or brand image because they are anonymous, shared on (trusted) forums and in the public view of potential visitors. Victims are not defenceless, though, as Tam's study shows, and may lower the value-destruction effects of a trolling episode by employing one, or a combination of three approaches—ignoration, rebuttal and re-orientation.

Orea-Giner, De-Pablos-Heredero and Vacas-Guerrero insist on the importance of studying value co-creation and co-destruction in the context of the information, communication, and automation revolution that is leading to the so-called Industry 4.0 era. Particularly, they focus on online word-of-moth (eWOM) communication to understand museum attributes that are relevant to visitors and promote co-creation during the whole visitors' journey. Unsurprisingly, the collection is seen as the most important core attribute of the analyzed museum. Compared to existing literature, though, the study's results uncover the importance of Industry 4.0 attributes in peripheral services such as an App and a Website. Interestingly, though, non-Industry 4.0 attributes such as toilets, a luggage storage, and accessibility are also enumerated among the new findings. While value may be co-created through enhancing these services, the study also discovered some contested attributes (such as the ticket price and the museum's identity that some locals perceive as extraneous) that could lead to value co-destruction.

Interpersonal encounters and social interaction

Glyptou's contribution investigates the case of Lesvos, a Greek destination in a sustained crisis since the arrival of the first refugees and immigrants in 2012. Research tends to concentrate on a crisis' peak leaving almost unexplored the tourists' response and engagement with the process of image co-creation for destinations in sustained crisis. In an holistic model, Glyptou proposes that tourism engagement in the image co-creation of such destinations is dependent on the pre-visit and during visit destination image, and the overall quality and intensity of interactions tourists had at destination with other tourists, local, and refugees, as well as their inclinations to create and share content on social media. Findings are very rich. Among these, we would like to point to the role of pre-visit perception of the host community and to interpersonal encounters at the destination. A positive affective image of the host-community trumps tourists' concern for their safety and induces them to visit the destination—an essential first step for the recovery of a destination's image. Encounters during the visit were the most significant variable directly related to image co-creation. In the light of the present crisis, one may suggest that host-communities that are known for being particularly welcoming and hospitable will be the first to recover.

Xing Han, Carolus Praet and Liyong Wang open a window on the Chinese outbound tourists and how social interaction with previously unknown other tourists shapes their perception of value co-creation and co-destruction. Whilst the direct or indirect presence of other tourists impacts on the tourism experience has been well discussed, the impact has been mainly framed as conductive of vale co-creation. However, Han et al. findings imply that value co-destruction is a real danger particularly in indirect interactions (those interactions that occur without explicit verbal on non-verbal communication between tourists). Negatively perceived encounters mostly occurred when the environment was considered to be crowded, reflecting the negative impact of over-tourism not only on the host community but also—though paradoxically—on the tourists themselves. Counterintuitively, the presence of tourists of the same nationality strengthened respondents' perception that value (in this case the authenticity of the place) is destroyed. The high incidence of indirect interaction leads the authors to conclude that, although the majority of their respondents were traveling with family or friends, they still tended to remain in their own 'bubble' and limit their over interaction with others. Whether this is characteristic of (specific segments of) Chinese outbound tourists or a stage in the coming-of-age of Chinese outbound tourism is subject for future research.

Designing and delivering co-created experiences

The study by Buxton and Michopoulou addresses both the role of social interaction as well as the role of process customization in value co-creation. It focuses on an unresearched leisure setting, spas, because they are often considered destinations in themselves and they are also inextricably linked to the pursuit of health and wellbeing—one of the most prominent forms of wellness tourism. It offers a welcome in-depth theoretical analysis of value co-creation and co-destruction concepts including value, authenticity, emotions and memorability of experiences, before showing how these can be applied to complex servicescapes and sensory-rich environments such as spas.

The fact that the design and delivery of co-created experiences entail several challenges is well shown in the article by Bezova and Azara. The first point that the authors wish to make is that value is co-created (rather than co-produced) when the customer is an active participant of the value creation process across all stages of the customer journey and that this process, for a series of reasons varying from the number of stakeholders involved to the lack of proper financing, may be particularly challenging for heritage sites. They probed this assumption by focusing on UK heritage sites and collecting insights from visitor experience managers. Findings suggest that, as in all co-creation processes, the characteristics, knowledge, and expectations of both parties highly determine the degree of co-creation that is possible. Interestingly, while respondents first point to a lack of engagement and demand from the customers, they also need to admit being better acquainted with the design and delivery of co-produced and customized experiences than co-created ones. One is therefore left wondering, whether the alleged visitors' profile and demands are the cause or the effect of the traditional (more co-production than co-creation) way in which heritage's sites shape their interaction with the customer. This article undoubtedly constitutes a good start for a much-needed reflection on the need to and practicalities of integrating co-created experiences in heritage site visits.

The paper by Michopoulou, Al-Qasmi, and Melpignano studies the role of desert tourists and camp managers in the process of value co-creation. It provides a comprehensive review of issues impacting co-creation as perceived by these two stakeholders, by connecting well-discussed concepts in the tourism literature such as place attachment and authenticity to the specific (and understudied) desert tourism context. Although the context and approach to this study is highly different from the UK study by Bezova and Azara, at least one of the conclusions is similar: the difficulty for service-providers on location (being guides on a heritage site or camp managers) to engage tourists in an effort of co-creation all along the customer's journey. A second similarity between the two studies may be found in the limitations imposed by the knowledge of (one of the) parties on the possibilities for co-creation. In this case, it is the camp managers' gaze, their understanding of guests' search for authenticity and their familiarity with social media that need to be challenged in order to come to a design and delivery process fully conducive to co-created experiences.

In conclusion, the examination of processes of value co-creation and co-destruction within tourism is now more pertinent than ever. Given the effects of constant sociocultural and environmental change and pandemic, and the huge challenges facing the sector, it is now more important than ever to understand what value is and how it can be created or destroyed. At the same time, findings presented in these papers point to the need of investigating additional factors that contribute to value co-creation in the context of tourism, but perhaps more importantly, factors that can result in value co-destruction. This issue continues the discussion on value co-creation and co-destruction, recognizing the variety of tourism servicescapes and experiences and emphasizing the necessity to examine tourism from beyond the narrow customer or provider angle; but rather focus on a wider spectrum of actors across value and distribution chains. This line of research is valuable therefore not just to expand this growing body of knowledge on a theoretical level; but to offer practitioners the critical insights required to rethink value within the design and delivery of their services, through the recovery stage and beyond.

Elena Cavagnaro

Eleni Michopoulou

Nikolaos Pappas

Revisiting Online Tourism Forums as Vehicles for Value Co-destruction

Aaron Tham and Mingzhong Wang

ABSTRACT

This paper revisits the value of online tourism forums as a vehicle for value co-destruction and corresponding responses by different users. Employing a content analysis of 469 trolling episodes recorded on Lonely Planet Thorn Tree forums, the research showed how such incidents allude to value co-destruction. However, the research also showcases how trolling victims respond to these unprovoked episodes by employing one, or a combination of three approaches—ignoration, rebuttal and re-orientation to deal with value co-destruction. In doing so, the research has contributed to a nuanced understanding of value co-destruction triggered by trolling, and corresponding responses as within forums. This provides a useful starting block for destination social media accounts to ensure that they take proactive measures to address trolling before they can create toxic online environments.

Introduction

Social media have radically transformed the tourism landscape, leveraging on the ubiquitous growth of internet connectivity, affordability and browsing speeds (Kristensen, 2013; Lamberton & Stephen, 2016). Proponents of social media in tourism have argued how they are powerful mechanisms to create a multiplicity of benefits to products and experiences in the form of creating awareness, stimulating interest, influencing purchase decisions and as a marker of service quality (Alonso et al., 2013; Ballantyne & Nilsson, 2017; Barrett et al., 2015). However, the ease of social media production and dissemination have created some challenges for businesses and tourists. Fake reviews, as well as incentivised electronic word of mouth, have cast light on the authenticity and reliability of user-generated contents (Choi et al., 2017; Filieri, 2016).

While there is a growing body of literature surrounding motivations and uses of social media (see for instance Leung & Bai, 2013; Munar & Jacobsen, 2014), there remains very little known about how the interactions with unknown others trigger trolling episodes, and the deleterious effects of such incidents on users. For the purpose of this research, a troll is defined as an individual who creates seemingly innocent contents within a digital landscape to obstruct, shame, provoke or antagonise other users, mostly on forum threads (Graham, 2019). Specifically, this research targets trolling episodes on tourism forums and locates how these incidents have altered the value co-creation/co-destruction landscape for a famous tourism brand, *Lonely Planet*. To the authors' best knowledge, only Mkono (2018) and Tham and Wang (2017) had undertaken some empirical study of trolling within tourism. This offers a point of departure in terms of understanding value co-destruction within forums as a dimension of the "dark side" of online deviant behaviour, especially when they were originally created for the benefit of information seeking and exchange of travel and tourism ideas (Colladon et al., 2019; Oriarde & Robinson, 2019). Whilst such studies have provided some exemplars of trolling, these authors conceded that further studies are necessary to understand the impacts of such online deviant behaviour and corresponding responses in a social media environment. This research seeks to unpack this gap in knowledge.

Literature review

The literature review will draw from the existing landscape of deviant behaviour to synthesise the current state of literature related to key work surrounding the concept of trolling within the overarching framework of value co-creation and destruction. Each of the subsequent sections will systematically illustrate extant knowledge to date and help present an informed setting for the research.

Value co-creation and co-destruction

Broadly defined, value is the derived outcome from an exchange between the gains received and the resources incurred in the transaction (Gallarza & Gil, 2008). In contemporary tourism environments, value is co-created from the exchange of ideas and actions between experience providers and tourists to ensure that there are mutually agreeable outcomes from the transactions (Camilleri, 2016; Cannas, 2018; Thomas-Francois et al., 2018). Tourists derive value from undertaking tourist activities to realise leisurely pursuits (Komppula, 2005; Sanchez et al., 2006). From a provider's perspective, the co-creation of value delivers a tourism experience that is of mutual benefit, which can lead to brand loyalty and returns on investment brand awareness and retaining loyal customers (Ellis & Rossman, 2008; Ritchie, 1999).

In comparison to value co-creation, value co-destruction (abbreviated as VCD hereafter) has received scant attention in scholarly literature. This may be attributed to the skewed perception that co-creation is the panacea for organisations to gain a competitive advantage. However, this narrow perspective has been challenged by a few scholars, who articulated that in some occasions, value is instead eroded as a result of the interactions between various stakeholder groups (Cova & Paranque, 2012; Daunt & Harris, 2017; Prior & Marcos-Cuevas, 2016). In the extant literature, VCD has occurred due to a few factors such as differing expectations of the service engagement, shifting attitudes as a result of the interactions and the erosion of trust due to other external stimuli (e.g. price competitiveness). VCD has also been slowly emerging within tourism as scholars pay greater attention as to how this has manifested, and its consequences to the sector. Table 1 illustrates a chronological order of the current body of knowledge concerning VCD in tourism.

From Table 1, studies to date have largely focused on actors, or participants in a physical service setting, with some within experimental designs or analysis of online reviews.

Table 1. Value co-destruction literature in tourism-related contexts.

Year	Author(s)	Context	Sample	Method	Findings	Gaps
2011	Echeverri and Skalen	Interviews	55 stakeholders of public transport in Sweden	Qualitative analysis	Value co-destruction from customers' ignorance of protocols, mismatch of greeting expectations, rigid applications of service roles	Observe interactions between providers and users B2B or e-commerce contexts
2014	Stieler, Weismann, and Germelmann	Silent protests at sports stadiums	40 fans	Interviews	Value co-destruction was an outcome of two scenarios: Silent protests eroded the value of attending live sports events Silent protests restrained fans from supporting their own teams	Long term effects of value co-destruction Other techniques rather than merely interviews Various contexts such as riots, pitch invasions etc Other stakeholders e.g. athletes, sponsors, TV audiences
2016	Gohary, Hamzelu, and Pourazizi	Service failure	254 students	Experimental design	Value co-destruction can lead to irreparable brand damage and significant negative word of mouth	Less service intensive engagements Qualitative and continuous analysis
2017	Camilleri and Neuhofer	AirBnB	850 reviews	Content analysis	Value co-destruction emerged as a consequence of negative emotions to the experience or country, lack of interactions/communication, absence of recommendations and poor etiquette	Other geographical or cultural contexts Interviews and observations Various sharing economy practices
2017	Quach and Thaichon	Luxury brand consumption	24 residents and tourists in Thailand	Interviews	Online platforms de-sensitive the human touch Exclusiveness is eroded due to the ease of creating online content featuring luxury brands Lack of informational cues online limits the ability to pre-test the item Online platforms do not facilitate personalised services	Other contexts of tourism and hospitality Male perspectives Online and offline environments Emerging markets
2018	Jarvi, Kahkonen, and Torvinen	Organisations in Finland	17 organisations	Interviews	Eight factors lead to value co-destruction: Absence of information, lack of trust, mistakes, inability to serve, not wanting to change, lack of clear expectations, customer misbehaviour and blame	Public sector insights Global perspectives Customer perceptions
2018			13	Interviews		Real-time environments

(*Continued*)

Table 1. Continued.

Year	Author(s)	Context	Sample	Method	Findings	Gaps
	Malone, McKechnie and Tynan	Ethical worldviews of participants			Negative emotions triggered incongruent outcomes and led to diminishing interests in some tourism destinations/experiences	
2018	Sthapit	TripAdvisor reviews	165 neutral reviews of 2 Finnish cruise lines	Content analysis/ Netnography	Cruise ship cabins and buffet contributed to no value creation outcomes	Refinement of deviant online behaviour Contexts other than hotel reviews Settings other than TripAdvisor
2019	Baker and Kim	Exaggerated online reviews	80 CIT and 90 experimental design outcomes	Critical incident technique and experimental design	Exaggerated reviews are strongly influenced by language complexity and emotions	
2019	Dolan, Seo and Kemper	Facebook complaints to airlines	1509	Content analysis	Value co-destruction occurs when solutions are incompatible/incongruent, support from others is misaligned, and when corporate responses are a mismatch with expectations	Other types of social media platforms Antisocial or deviant behaviours
2019	Gkritzali, Mavragani, and Gritzalis	Twitter tweets on financial crisis in Athens	93,300	Sentiment analysis using Neuro-Linguistic Programming	Twitter is a tool that can exacerbate negative eWOM, leading to a less than appealing destination image	Other forms of crises Different contexts
2019	Kim, Byon, and Baek	Golf tournament spectators	318	SEM	Dysfunctional behaviour erodes emotional value, resulting in co-destructive outcomes	Other sports Cross cultural differences
2020	Lund, Scarles, and Cohen	Social media of Visit Denmark	Storytelling in the area of report, presentations and website	Interviews	Storytelling can rectify value co-destruction occurring on social media	
2019	Luo, Wong, King, Liu, and Huang	Visitors to Shanghai Disneyland	765	Questionnaires	Value co-destruction within C2C encounters impact on users' brand affection and allegiance C2C value co-destruction also dilutes service quality outcomes	Consider different cultural contexts Longitudinal studies
2019	Sthapit	AirBnB negative reviews on TrustPilot	694	Content analysis	Bad behaviour of hosts and poor customer service from AirBnB led to value co-destruction	
2019	Sthapit and Bjork	Reviews of Uber	75	Netnography	Value co-destruction was triggered by drivers' bad behaviour and poor customer service	Discussions limited to those stating poor experiences using three keywords
2019	Sthapit and Jimenez-Barreto	Interviews with AirBnB users	21 users in Spain and Finland	Qualitative analysis	Lack of responsiveness was a facet of value co-destruction	

2019	Yin, Qian, and Shen	Tweets on Sina Weibo	8813	Thematic content analysis	Maintenance and vandalism issues trigger value co-destruction, leading to negative eWOM	More extensive social media dataset Supplemented by other methods e.g. interviews, ethnography Cross cultural insights
2020	Jarvi, Keranen, Ritala, and Vilko	Interviews, reflective diaries and TripAdvisor reviews	5, 15 and 344 respectively	Qualitative analysis	Value co-destruction originates from providers' inability to provide service, context rigidity and incoherent marketing, while customers are from excessive expectations, insufficient communication and inappropriate behaviour	Other contexts rather than hotels Longitudinal design Managerial implications
2020	Yeh, Fotiadis, Chiang, Ho, and Huan	Questionnaires from tourists with hotel experiences in Taiwan	601	Structured equation modelling	Two mediators of value co-destruction are desire for revenge and desire for recovery. Negative emotions translate to revenge intentions as a means of getting back procedural justice, though service recovery is welcomed.	Different generations Geographical contexts Longitudinal data

There remain significant knowledge gaps on VCD, especially in the space of online interactions that have become ubiquitous in tourism. Furthermore, social media offer the proximity of users to be associated with strangers for a plethora of tourism-related regions. The lack of knowledge surrounding VCD in such facets of engagement is the basis to undertake this research devoted to deviant behaviour as framed through online forums.

Deviant behaviour

The term deviant behaviour may be understood as dysfunctional words or actions that are incongruent to the expected setting or culture (Akers, 1968). In tourism, deviant behaviour can take various forms, such as vandalism (Bhati & Pearce, 2016) and drug parties (Uriely & Belhassen, 2006). For instance, destinations such as Chersonnissos, Greece, offer the millennial market an opportunity to escape from their usual environments to participate in deviant behaviour such as binge drinking practices due to low prices of alcohol, where such behaviour is likely to be frowned upon in their home countries (O'Connell, 2009). The consequence of deviant behaviour can likewise result in several consequences. A tourist may have to face time in jail, pay a fine, or be deported to his or her own country of origin. Regardless of the severity of the deviant act, it is likely that such incidents can generate negative and unwanted publicity for both the origin and the destination (Ravescroft & Gilchrist, 2009). For this reason, a two-prong approach towards tourist behaviour has been proposed. First, some countries have attempted to educate their citizens to behave in a civilised manner and follow local customs and laws when travelling abroad (Waldmeir, 2015). A second approach is to use warnings and reminders that deviant behaviour will result in punitive measures (Starmer-Smith, 2009). Taken in combination, these approaches are aimed at mitigating the likelihood, severity and frequency of deviant behaviour during travel.

Despite the good intentions of home and destination countries, deviant behaviour can still be triggered by a range of factors. Some scholars believe that peer pressure is a key antecedent of deviant behaviour. Using the case of rioting during Schoolies Week on the Gold Coast in Australia, Lubman et al. (2014) found that the deviant behaviour was prompted by a combination of alcohol or drug-induced actions under the influence of peers. Another factor prompting deviant behaviour is the perception that one can misbehave as tourism is outside one's usual environment (Spencer & Bean, 2017). Finally, deviant behaviour may be a result of wanting to show off one's acquisitions during travel. Recent incidents where tourists have climbed onto expensive relics in order to capture themselves in selfies and in this process, destroying the monuments have reared the ugly head of narcissism (Loughrey, 2016; Squires, 2015). Clearly, deviant behaviour in tourism can occur for a variety of reasons and are highly unpredictable.

Trolling and online deviant behaviour

The focus of this paper is on trolling as a form of online deviant behaviour. In contrast to the previous section, online deviant behaviour is more likely to occur at the pre-consumption stage of the user experience (Cook et al., 2018; Jenks, 2019). In a tourism setting, the advent of social media to aid in tourism planning and decision-making has compressed the time and effort to acquire required information (Munar & Jacobsen, 2013). As such, it is highly probable that trolling is encountered during this timeframe. However, there remains a paucity of literature surrounding the impact of trolling and how it alters the hedonic pursuits of tourism experiences (Mkono, 2018; Tham & Wang, 2017).

Trolling is a subset of cyberbullying (Coles & West, 2016). Existing literature has suggested that cyberbullying takes the form of action or words that attack individuals during their digital engagements (Brietsohl et al., 2018; Slonje et al., 2013). Cyberbullying can occur where both the perpetrator and the victim are known to each other. For example, studies have revealed that cyberbullying may occur in a school environment, where students may get picked on because of their physical attributes and labelled unkind words circulated among peers in cyberspace (Hinduja & Patchin, 2013; Huang & Chou, 2010; Mishna et al., 2010; Smith et al., 2008). Cyberbullying can likewise occur in a work environment (D'Cruz & Noronha, 2013; Piotrowski, 2012; Snyman & Loh, 2015). For instance, an employee is given a reprimand for being late, and the naming and shaming process is sent to all staff via an email. Cases of cyberbullying can result across a spectrum of contexts but are perhaps harder to mitigate given the ease in which synchronous transmissions occur (Hoff & Mitchell, 2009).

Whilst sharing some similarities in possessing cyberbullying characteristics, trolling is perhaps more difficult to detect because a troll can surprise any online user and antagonise one's feelings and emotions (March, 2019). Moreover, the antidote for trolling at present is to kick out the perpetrator from the social media site, such as forums (Dynel, 2016). However, with the ease of setting up a new account, the troll can re-enter the forum and continue to attack others in a similar fashion (Synnott et al., 2017).

Literature also suggests that trolling is not limited to user profiles or geographical regions (Elmezeny et al., 2018). However, Hong and Cheng (2018) postulated that university students appear most susceptible to be on the receiving end of trolling attacks due to their disposition to post more often on social media sites. Case and King (2018) likewise concurred that university students are more likely to be victims of trolling, where their study found that the greater the length of time spent online, the higher the probability that such a sample would fall prey to trolls.

The prevalence of trolling has not gone unnoticed, with literature more recently emerging to depict trolling characteristics and motivations for such online deviant behaviour (see Escartin, 2015; Nicholls & Rice, 2017; Patterson et al., 2017; Sanfilippo et al., 2017). For example, Bishop (2014) classified trolling into four types:

(1) Haters—individuals who provoke others for no apparent reason
(2) Lolcows—individuals who antagonise others so as to draw attention to themselves
(3) Bzzzters—talkative individuals who disregard fact or value of contributions
(4) Eyeballs—individuals who wait for a specific time to infuriate others

Yet, very little is known about trolling in given contexts (see Cruz et al., 2018; March & Marrington, 2019), such as tourism. Derived from the gaps in knowledge are the research questions:

How are trolling episodes on tourism forums a vehicle for value co-destruction (VCD)?

How do forum users respond to trolling episodes?

Methodology

This research applied a content analysis approach based on free-flowing trolling episodes manifested in publicly available discussion threads within the Lonely Planet Thorn Tree forum (https://www.lonelyplanet.com/thorntree/welcome) between January 2012 and June 2017. Thorn Tree is the online forum platform of the popular travel guidebook *Lonely Planet* (https://www.lonelyplanet.com/thorntree/welcome). Thorn Tree was developed as a vehicle to further extend the organisation's reach as it confronts a challenging environment where guidebook sales have continually decreased amidst the ease in which tourism recommendations can be procured electronically (Robin, 2013). The slowdown in print media has had major consequences on Lonely Planet, especially when it shut several global offices and retrenched staff because of diminishing revenues (Clune, 2013). While several of its titles are now available as an e-book, social media have taken over the digital media landscape, and Thorn Tree offers Lonely Planet an opportunity to stay competitive (Butler & Paris, 2016). Forums such as Thorn Tree have been utilised as frames of investigation as evidenced in other tourism studies due to its advantage of presenting data in an unobtrusive manner (e.g. Hwang et al., 2013; Paris & Teye, 2010; Pirolli, 2018). Guided by extant literature, Thorn Tree (abbreviated as TT thereafter) forums presented a credible base to justify their inclusion for the purpose of this research.

A web crawler was developed using Python programming language to retrieve threads published before June 2017 that featured the term "troll" within TT pages. This research only considered threads with the term "troll" in them even though a thread without such a term may have actual trolling behaviours. This was due to two reasons. First, it is impractical to manually read every post in the forum to decide if it is a troll thread based on its semantics. Computer programs are still not 100% accurate and reliable for semantic understanding and content analysis to automate this process. Second, the fraction of trolling threads without the term "troll" is neglectable because TT has active user participations in discussion and veteran users are keen to tag the thread as "troll" if they feel so. We have manually investigated more than 100 random threads without the term "troll" and found none of them have trolling behaviours. Therefore, it is argued that the proposed data retrieval method has neglectable probability to miss an accurate depiction of a troll. The web crawler collated 787 threads that possessed the term troll and its variants, such as trolling. There were no threads containing the term "troll" before January 2012. The initial 60 trolling episodes were independently coded by the researchers to ascertain what had triggered such instances, and how others in the thread had responded.

Trolling episodes were informed by extant literature that provided an indicative typology of such online deviant behaviour (Bishop, 2014; Hopkinson, 2013; Kavanagh et al., 2016). Knowing how to categorise trolling episodes was instrumental to this research, as this would suggest more informed approaches to address such phenomena. The researchers concurred on how to categorise 46 of these episodes, with the remaining 14 agreed upon following subsequent discussions. Subsequently, the rest of the dataset was assessed to determine if these were reflective of trolling. Instances where the word "troll" appeared without any elaboration, or where the term was part of other contexts, such as destinations such as *Troll's Tongue* were excluded from further analysis. This resulted in 469 trolling episodes employed for the research.

Results and findings

This research aimed to gain insights into the trolling episodes within online tourism forums, and in turn explore how these have manifested to VCD. This is important especially in the context of tourism when travel planning often conjures images of pleasure and relaxation.

Origins of trolling episodes

There was an almost equal spread in terms of the origins of trolling episodes. 239 out of the 469 episodes (50.9%) were associated with the original posters being labelled as trolls, while the remaining 230 (49.1%) of these deviant online behaviours were instigated by other users within the threads. This finding provides a fresh perspective to examine trolling manifestations, as previous literature appeared to put most of the blame on other users attacking "innocent" original posters (Lumsden & Morgan, 2017). However, the almost even split to suggest that original posters are equally culpable of trolling behaviour reveals the practice of soliciting for hostile responses through the process of "baiting". According to Golf-Papez and Veer (2017), baiting is the use of text or images that seek to provoke others to respond in an antagonising manner because of the way in which a viewpoint has been phrased. This was evident in several trolling encounters collated in the data, especially when the original post appeared to lure others towards a specific stance, as exemplified in the following exchange on TT:

> Hello. I am from America so I was wondering what countries are more accepting … I was told to avoid Spain, Russia, Australia and Germany since I am black … I don't want to go to places where blacks are majority, I want some diversity or something different … I know Europe is not fond on their African immigrants … so is it true England is likely the next most tolerant place outside of America? (Original poster)
>
> If you begin with the attitude "people will be mean to me because I am black, American etc." you will probably feel that anywhere. Nothing except changing your attitude will ever change that perception. (Other user response 1)
>
> There are a number of issues assumptions and statements in your posting. So many, in fact, that I'm wondering if you're a troll …. (Other user response 2)

Another example of baiting in the context of the forums was obtained from an alternative thread featuring a complaint about poor service in Cambodia:

> I'm traveling on a budget and when I saw I could get a bed for $1 a night at utopia I was very excited. I arrived and checked in … I go and drink at the bar a bit, it's quite lively and fun, a good backpacker scene and happy hour beers for $0.25 … I returned a bit later and went to order another beer for $0.25. The bar tender proceeded to come out from behind the bar and tried to physically force me off the bar by grabbing my legs etc … I went to immediately check out, feeling my safety was jeopardised. I highly recommend you say away from this place and fork up an extra $1 or $2 to find other accommodation. (Original poster)
>
> What's the saying? You pay peanuts, you get monkeys. (Other user response 4)
>
> Getting a $ 2.50 room and then demanding .25cts beers all night is a bit strong even by cheapo backpacker standards. Right, I would have turfed him out, too (sans the violence). (Other user response 6)

In this incident, the baiting resulted from the original poster's demands for low prices at the accommodation and bar for the entire night of stay, which seemed to provoke responses to the unrealistic expectations held by the original poster. In this case, the complaint of the service could have resulted in others taking sides with the original poster, or as shown, created a backlash as to the tourist's behaviour. While there was very little authentication as to whether such an incident did take place, other users on the thread were quick to point out that the original poster should take responsibility for such an outcome and was justifiably kicked out of the accommodation and bar.

One of the common characteristics witnessed in the forums that could be deemed antagonising is the use of the pronoun "you". In the quotes, the pronoun is often undertaken as a finger-pointing exercise to accuse the original poster that border on one's sense of entitlement and tourist expectations. These incompatible differences trigger antagonistic outcomes as manifested across online forums, thereby eroding the purpose of the platform as a medium of information exchange and solicitation (Dolan et al., 2016). As the conversations above indicated, the process of baiting can traverse boundaries beyond what TT was intended for. This is especially the case when the exchanges go off-topic and infringe topics such as ethnicity, religion and sexuality, which go against the ethos of TT. Hence, baiting practices by both original posters and other users are antecedents to trolling episodes. The implication of baiting to online forums necessitates the urgent need to establish behavioural norms within social media, as the ease of publicising information, often without repercussions, can deter potential members from further participation. Forums should continuously remind users of what constitutes acceptable on-site behaviour, and then run the risk of being expelled should recalcitrant behaviour occur.

Responses to forum engagements

Three common responses emerged from the findings based on users' engagement with TT. The three responses are ignoration, rebuttal and re-orientation. As shown in Figure 1, 146 out of the 469 episodes (31.1%) were in the category of ignoration, 177 (37.7%) were rebuttal, 83 (17.7%) were re-orientation.

In addition, 63 episodes (13.4%) could not be classified for two main reasons:

- One, the original poster was considered to be the troll but left the conversation after the first post, or
- Two, most posts were removed due to moderation.

Each of these responses will be subsequently discussed to show that responses to trolling are heterogenous, which differs to what other scholars have found (Elmezeny et al., 2018). Instead, our findings validate the assertions of Hardaker (2015), in that user responses often differ.

Ignoration was one of the strategies employed by forum users on TT when encountering trolls. The word "ignore" was used to advice original posters to disregard troll comments, as the following quotes indicate:

> Ignore E. He is a troll and spends his time making similar comments on other threads. He has control issues and is very protective over lonely planet thorntree. He needs to get a life.

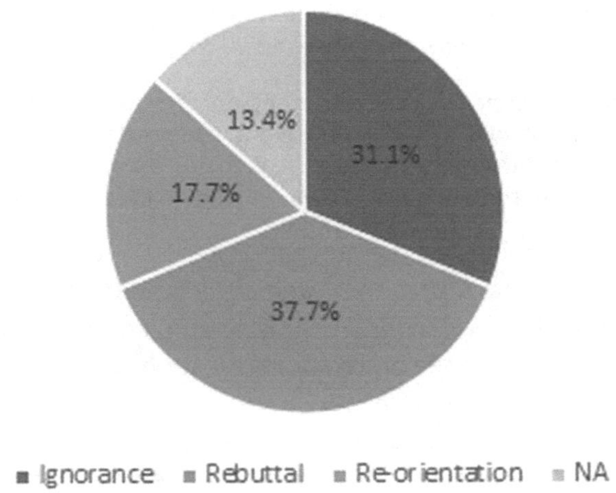

Figure 1. Distribution of response categories.

> Thank you L and V for replying to my question and ignoring the troll in the end.

> #3 is a troll … ignore.

> Folks please don't engage the troll under the bridge. If he is ignored, he will be left to talk to himself …

Employing ignoration as a tactic to thwart trolls has been suggested elsewhere (Cook et al., 2018; Fichman & Sanfilippo, 2015; Hardaker, 2015; Jenks, 2019). Furthermore, if we accept that trolling is a form of conflict, then by extension, ignoration has also been adopted as a strategy for dispute resolution in other contexts (Deutsch, 1990; Jordan & Troth, 2004; Leung, 1988; Xie et al., 1998). While such strategies to avoid conflict may be easily employed, other scholars contended that these are short-term strategies and the conflict is yet to be resolved in the long-term (Richardson, 1995; Tjosvold, 2008). Trolls can still attack their victims in the same forum, or elsewhere in circumstances of temporal ignoration. This provides a fresh conceptualisation of how ignoration is a proxy to VCD, and extends Echeverri and Skalen (2011)'s work to an online space as evidenced in this research.

Rebuttal was another popular strategy evident from the findings. Across the TT threads where trolling was encountered, rebuttal took on a different response mechanism as victims sought to engage with trolls to reason or persuade trolls to accept their perspective. Some of the examples of rebuttal in action are provided in these subsequent quotes:

> Why do people always think I'm a troll? Yes I know, everything I do is always such a horrible idea, whatever you guys not everyone lives as conventionally as you do … Personally, I'm naturally interested in everybody's life story, and when I come into their little-travelled town with a backpack, it gets them interested, and it's a lot easier to forge friendships when there's mutual interest … I'm inclined to get defensive when my lifestyle is called heinous, but actually, I think people just don't understand alternative lifestyles so sorry.

> There are those who have posted on this thread who, I believe, troll threads on this and other forums looking to respond to questions. They don't give an answer, but respond with " … gee, I hope someone will help you." I was using this forum as a last resort thinking … I

> only wanted those who knew something about getting a resident visa to respond … This forum like others, if you don't know the answer, it would be best not to respond at all.
>
> #13—Thank you for your response. I can see your point and apologise for the fact that this post seemed frivolous or idle … The reason I got annoyed is the fact I was called a "troll"—something I definitely am not … If the offending post is a "troll" post for time-wasting or whatever, I can see why since I am not intending to do this trip anytime soon …

In the three unrelated rebuttals, the original posters take a defensive stance against being call a troll, on the basis that their primary intentions were to clarify perhaps obscure tourism information within TT. For this reason, rebuttals are a demonstration of the original posters' rights to seek relevant information, and therefore being called a troll was an unjust outcome. Put in perspective, the social norms of information seeking on TT appear to be implicit as to what questions are asked, or how these have been framed (either positively/negatively). Hence, new users to TT can be unfairly labelled trolls when they are not acquainted with the "rules of engagement". Problematising trolling incidents within online forums is the deceptive simplicity of written prose that marginalises embedded contexts associated with the information seeking behaviour (Gully, 2012). While it is acknowledged that adopting difference stances is quite common on public forums, entering argumentative discourses is a proxy to persuade others to stand in solidarity with one's perspectives (Chandrasegaran & Kong, 2006). However, at a cursory level, written exchanges on any online forum do not reveal possible emotional and affective attachments one might have towards the need for rebuttal (Wu, 2015). The findings instead reveal rebuttal efforts among the forum users have triggered vitriol, or incensed users as a dimension of VCD. Therefore, in such instances, unclear expectations, or implied contexts, can lead to miscommunication, or worse, incidents of trolling on forums (Dixon, 2014; Hardaker, 2010). This, therefore, manifests another dimension of VCD within tourism (Clinnin & Manthey, 2019; Massanari, 2019; Mkono, 2018). Exemplars of comments that suggest re-orientation outcomes are as follows:

> Hello! I'm planning a trip to Cambodia in early December for 18ish days and after doing some research (and reading lots of posts here!) I have this itinerary. What do you guys think? … Thanks for your help/recommendations! (Original poster, first message in thread)
>
> …
>
> It sounds like we are missing Battambang and Kratie so I'll figure out how to fit those in! (Original poster, halfway through thread)
>
> …
>
> Oh please. Surely you have something better to do than troll travel forums making fun of other people's itineraries and refusing to share your 17 years of knowledge! (Original poster, in response to troll)
>
> …
>
> … we won't be staying in Koh Kong the city but probably one of the ecolodge type places in the area. (Original poster, re-orientating accommodation choices)

For users who have decided to re-orientate their choices or preferences following TT troll encounters, there was enough influence to induce the need for change. This outcome is a contribution to literature as it is essential to ascertain conditions in which such circumstances have occurred, and why they have transpired. In the three exemplars above-mentioned, re-orientation precipitated from significant engagements and contributions within the forum. There were attempts to ignore irrelevant posts, rebuttals based on pre-held notions of how one should travel or conduct oneself, but escalated to persuasion in order to effect influence, leading to re-orientation of original posts. This escalating effect of forum engagement is perhaps moderated by time and effort to be involved and cautiously interact with others in the thread. As other studies have alluded to, online forums are akin to groups that go through different stages of forming, storming, norming and performing (Dietz-Uhler et al., 2005; Fayard & DeSanctis, 2005; Johnson, 2001; Nisbet, 2004). As such, re-orientation is triggered by greater involvement in online forums and the susceptibility to influence. While re-orientation is not necessarily attributed to any given troll, there is some evidence that the benefits of TT outweigh the psychological costs of encountering trolls. Compared to extant literature, this outcome differs to the nomenclature of VCD responses as suggested by Yeh et al. (2020). Their work showed how desire for revenge or recovery were mediators towards disseminating negative word of mouth, which is valuable in scope for a hotel context defending its reputational interests and brand. However, this research illuminates that re-orientation is a further response to VCD as it differs in the form of a forum site that does not have value co-creation measured in economic terms but carry much further weight in terms of psychological value outcomes. This brings a fresh perspective of VCD manifestation and responses to the paucity of literature on trolling in tourism.

Destination images imposed within social media

Tourism literature has concluded that destination images are conjured out of cognitive and emotional responses, explaining how decision-makers select locations that best match desired needs and interests (Karl et al., 2015; McCabe et al., 2016). Hence, attacks launched on TT forums by trolls are likely to provoke negative responses, as the following quotes illustrate:

> Hmmm … now I'm confused! If I read S's reply I feel reassured and if I read T's reply I feel like an irresponsible parent! But … obviously I don't want to put their safety at risk!

> Not sure how trying living in San Miguel is making "mistake after mistake?" And all I did was come on here and ask about health concerns in San Cristobal and had several people proceed to tell me what a fool I am … I don't need the drama. No more Latin America.

The quotes highlight incongruent destination images held by different individuals on the forums. The confrontations that have emerged out of trolling episodes reflect the desire and willingness of TT users to jostle and impose their destination images on others within the forum. While destination image can be highly personalised, calling someone a troll is thereby throwing accusations that his or her views are somewhat incompatible with their own, and thereby triggering animosity between users.

Unlike face to face conversations where destination choice outcomes can be influenced by word of mouth, the façade of an anonymous online screen causes much more damage as trolling originates from unknown persons, is enshrined in the forum, and in the public view of numerous others. Such behaviour can have the potential to erode one's destination image and thereby question one's possible choice of destinations (Mkono & Tribe, 2017). In this sense, the research offers a breakthrough in terms of uncovering insights to how trolling episodes create psychological tensions associated with destination image destruction, and as a consequence alter the terrain of choice sets within tourism (see Thai & Yuksel, 2017). While choice set literature espouses that evoked sets lead to eventual destination selection, trolling has a negative impact on this and can cast destinations to a reject set rapidly. Even though trolling has been documented over the past decade across a multiplicity of contexts, organisations have hardly responded to defray the psychological risks that online consumers are exposed to (Herring et al., 2002). This reveals another dimension of VCD emanating from trolling encounters.

Conclusion

In conclusion, this research set out to answer the research question that explored how online tourism forums are vehicles for VCD. These objectives have been realised on the back of investigating 469 TT forum threads where trolling episodes were documented. The research has stimulated a more nuanced conceptualisation of what appears to be a homogenous notion of a troll as someone who exhibits mischievous acts. Instead, the trolling episodes reveal that such disputes on social media occur across a range of factors, including attacks on the tourist, the destination or trip characteristics among other considerations. While individuals can choose to ignore or rebut the presence of trolls, destinations and other organisations cannot overlook this phenomenon. This is because more far-reaching implications that could have detrimental effect on brand and image are harder to recover from (Rauschnabel et al., 2016). Moreover, Linvill and Warren (2019) reported on the growing number of troll factories operating in locations, such as Russia. In these factories, individuals are paid to destroy and attack unsuspecting victims, and earn a living from performing online deviant acts.

The growth of trolling in light of increased online presence among virtual communities has hastened the developments of troll detection and mitigation software to address such deviant behaviour (Chiregi & Navimipour, 2016; Fornacciari et al., 2018; Paavola et al., 2016; Tayade et al., 2017). These tools employ machine learning techniques to mitigate the spread of trolling and can be customised to the mode of engagement and user profiles. Despite these efforts, detractors of mechanistic approaches are convinced that "soft" approaches, such as managing expectations are required (Lopes & Yu, 2017). The research indicated that tourists attempt to get quick and convenient responses by posing broad questions related to the pursuit of their proposed travel intentions to pre-selected destinations. Instead, some other participants perceive this to be an unprofessional act and hold expectations for tourists to at least have some preparation made prior to engaging on the social media platform offered by TT forums. This had then manifested into cognitive and emotional incongruence, which therefore leads to VCD, as alluded to by several scholars (Baker & Kim, 2019; Camilleri & Neuhofer, 2017; Kim et al., 2019; Malone et al., 2018). Clearly, there is a need to manage the service expectations given that TT embraces a highly interactive form of communication, involving peer-to-peer contents.

A final contribution of the paper is to illustrate that social media have increased the propensity for anyone to voice opinions online and imposing their views as potentially the right perspective. This is particularly the case for TT forums, as the research showed on numerous occasions how innocuous comments made from original posters triggered a variety of responses that stirred up trolling propensities on the back of differing destination images. While extant studies have showed that each individual's pre-held destination image can vary, social media have opened up a platform where such appraisals are becoming increasingly eroded through the exchange of views by unknown others. As such, there are times where forums like TT become battlegrounds, rather than upholding the ethos of its *communitas* (Baccarella et al., 2018).

Research limitations

Some limitations are identified in the conduct of this research. First, the data was obtained from a single tourism forum, which may limit its ability to generalise the findings. Second, the analysis was conducted using solely corpus text from the forums. Third, the research is conducted through the lens of English-speaking forum communities and has ignored other languages in their representations of trolling within tourism forums. Such limitations may be overcome in future by speaking with forum moderators to elicit insights as to their experiences in managing trolls, and their effectiveness. These limitations notwithstanding, the research has amplified how trolling is conceptualised within social media, leading to value co-destruction.

Future studies

This research has created avenues to pursue future studies. Subsequent streams pertaining to trolling could employ a mixed-method approach to triangulate the forum contents with other sources, such as questionnaires, interviews or focus groups. Other scholars may choose to also undertake longitudinal studies to ascertain if trolling episodes have diminished in terms of their frequency and severity (length of thread).

Overall, the research has expanded the implicit knowledge of trolling in terms of tourism development within the online domain. It illuminates that unless more stakeholders are involved to nip this online deviant behaviour, it can manifest and inflict psychological damage to the detriment of destinations, products and services. In an era where technology drives tourism behaviour, trolling can no longer be ignored as a destination's success hinges on their ability to attract potential tourists.

Disclosure statement

No potential conflict of interest was reported by the author(s).

ORCID

Aaron Tham http://orcid.org/0000-0003-1408-392X
Mingzhong Wang http://orcid.org/0000-0002-6533-8104

References

Akers, R. L. (1968). Problems in the sociology of deviance: Social definitions and behaviour. *Social Forces*, *46*(4), 455–465. https://doi.org/10.2307/2575380

Alonso, A. D., Bressan, A., O'Shea, M., & Krajsic, V. (2013). Website and social media usage implications for the further development of wine tourism, hospitality, and the wine sector. *Tourism Planning & Development*, *10*(3), 229–248. https://doi.org/10.1080/21568316.2012.747989

Baccarella, C. V., Wagner, T. F., Kietzmann, J. H., & McCarthy, I. P. (2018). Social media? It's serious! Understanding the dark side of social media. *European Management Journal*, *36*(4), 431–438. https://doi.org/10.1016/j.emj.2018.07.002

Baker, M. A., & Kim, K. (2019). Value destruction in exaggerated online reviews: The effects of emotion, language, and trustworthiness. *International Journal of Contemporary Hospitality Management*, *31*(4), 1956–1976. https://doi.org/10.1108/IJCHM-03-2018-0247

Ballantyne, D., & Nilsson, E. (2017). All that is solid melts into air: The servicescape in digital service space. *Journal of Services Marketing*, *31*(3), 226–235. https://doi.org/10.1108/JSM-03-2016-0115

Barrett, M., Davidson, E., Prabhu, J., & Vargo, S. L. (2015). Service innovation in the digital age: Key contributions and future directions. *MIS Quarterly*, *39*(1), 135–154. https://doi.org/10.25300/MISQ/2015/39:1.03

Bhati, A., & Pearce, P. (2016). Vandalism and tourism settings: An integrative review. *Tourism Management*, *57*, 91–105. https://doi.org/10.1016/j.tourman.2016.05.005

Bishop, J. (2014). Representations of 'trolls' in mass media communication: A review of media-texts and moral panics relating to 'internet trolling. *International Journal of Web Based Communities*, *10* (1), 7–24. https://doi.org/10.1504/IJWBC.2014.058384

Brietsohl, J., Roschk, H., & Feyertag, C. (2018). Consumer brand bullying behaviour in communities of service firms. In M. Bruhn & K. Hadwich (Eds.), *Service business development* (pp. 289–312). Springer Gabler.

Butler, G., & Paris, C. M. (2016). Reaching the end of the unbeaten path? An analysis of Lonely Planet's digital media "requiem". *Anatolia*, *27*(3), 377–388. https://doi.org/10.1080/13032917.2016.1191770

Camilleri, M. A. (2016). Responsible tourism that creates shared value among stakeholders. *Tourism Planning & Development*, *13*(2), 219–235. https://doi.org/10.1080/21568316.2015.1074100

Camilleri, J., & Neuhofer, B. (2017). Value co-creation and co-destruction in the airbnb sharing economy. *International Journal of Contemporary Hospitality Management*, *29*(9), 2322–2340. https://doi.org/10.1108/IJCHM-09-2016-0492

Cannas, R. (2018). Diverse economies of collective value co-creation: The open monuments event. *Tourism Planning & Development*, *15*(5), 535–550. https://doi.org/10.1080/21568316.2018.1505651

Case, C. J., & King, D. L. (2018). Internet trolling victimization: An empirical examination of incidence in undergraduate business students. *Research in Higher Education Journal*, *34*. https://files.eric.ed.gov/fulltext/EJ1178405.pdf

Chandrasegaran, A., & Kong, K. M. C. (2006). Stance-taking and stance-support in students' online forum discussion. *Linguistics and Education*, *17*(4), 374–390. https://doi.org/10.1016/j.linged.2007.01.003

Chiregi, M., & Navimipour, N. J. (2016). A new method for trust and reputation evaluation in the cloud environments using the recommendations of opinion leaders' entities and removing the effect of troll entities. *Computers in Human Behavior*, *60*, 280–292. https://doi.org/10.1016/j.chb.2016.02.029

Choi, S., Mattila, A. S., Van Hoof, H. B., & Quadri-Felitti, D. (2017). The role of power and incentives in inducing fake reviews in the tourism industry. *Journal of Travel Research*, *56*(8), 975–987. https://doi.org/10.1177/0047287516677168

Clinnin, K., & Manthey, K. (2019). How not to be a troll: Practicing rhetorical technofeminism in online comments. *Computers and Composition*, *51*, 31–42. https://doi.org/10.1016/j.compcom.2018.11.001

Clune, B. (2013). Is this goodbye to Lonely Planet? https://www.theguardian.com/commentisfree/2013/jul/22/lonely-planet-melbourne-staff

Coles, B. A., & West, M. (2016). Trolling the trolls: Online forum users constructions of the nature and properties of trolling. *Computers in Human Behavior*, *60*, 233–244. https://doi.org/10.1016/j.chb.2016.02.070

Colladon, A. F., Guardabascio, B., & Innarella, R. (2019). Using social network and semantic analysis to analyze online travel forums and forecast tourism demand. *Decision Support Systems*, *123*, Article 113075. https://doi.org/10.1016/j.dss.2019.113075

Cook, C., Schaafsma, J., & Antheunis, M. (2018). Under the bridge: An in-depth examination of online trolling in the gaming context. *New Media & Society*, *20*(9), 3323–3340. https://doi.org/10.1177/1461444817748578

Cova, B., & Paranque, B. (2012). Value creation versus destruction: The relationship between consumers, marketers and financiers. *Journal of Brand Management*, *20*(2), 147–158. https://doi.org/10.1057/bm.2012.46

Cruz, A. G. B., Seo, Y., & Rex, M. (2018). Trolling in online communities: A practice-based theoretical perspective. *The Information Society*, *34*(1), 15–26. https://doi.org/10.1080/01972243.2017.1391909

Daunt, K. L., & Harris, L. C. (2017). Consumer showrooming: Value co-destruction. *Journal of Retailing and Consumer Services*, *38*, 166–176. https://doi.org/10.1016/j.jretconser.2017.05.013

D'Cruz, P., & Noronha, E. (2013). Navigating the extended reach: Target experiences of cyberbullying at work. *Information and Organization*, *23*(4), 324–343. https://doi.org/10.1016/j.infoandorg.2013.09.001

Deutsch, M. (1990). Sixty years of conflict. *International Journal of Conflict Management*, *1*(3), 237–263. https://doi.org/10.1108/eb022682

Dietz-Uhler, B., Bishop-Clark, C., & Howard, E. (2005). Formation of and adherence to a self-disclosure norm in an online chat. *CyberPsychology & Behavior*, *8*(2), 114–120. https://doi.org/10.1089/cpb.2005.8.114

Dixon, K. (2014). Feminist online identity: Analyzing the presence of hashtag feminism. *Journal of Arts & Humanities*, *3*(7), 34–40. https://doi.org/10.18533/journal.v3i7.509

Dolan, R., Conduit, J., Fahy, J., & Goodman, S. (2016). Social media engagement behaviour: A uses and gratifications perspective. *Journal of Strategic Marketing*, *24*(3–4), 261–277. https://doi.org/10.1080/0965254X.2015.1095222

Dolan, R., Seo, Y., & Kemper, J. (2019). Complaining practices on social media in tourism: A value co-creation and co-destruction perspective. *Tourism Management*, *73*, 35–45. https://doi.org/10.1016/j.tourman.2019.01.017

Dynel, M. (2016). Trolling is not stupid: Internet trolling as the art of deception serving entertainment. *Intercultural Pragmatics*, *13*(3), 353–381. https://doi.org/10.1515/ip-2016-0015

Echeverri, P., & Skalen, P. (2011). Co-creation and co-destruction: A practice-theory based study of interactive value formation. *Marketing Theory*, *11*(3), 351–373. https://doi.org/10.1177/1470593111408181

Ellis, G. D., & Rossman, J. R. (2008). Creating value for participants through experience staging: Parks, recreation, and tourism in the experience industry. *Journal of Park and Recreation Administration*, *26*(4), 1–20.

Elmezeny, A., Wimmer, J., dos Santos, M. O., Orlova, E., Tribusean, I., & Antonova, A. (2018). Same but different: A comparative content analysis of trolling in Russian and Brazilian gaming imageboards. *Game Studies*, *18*(2). http://gamestudies.org/1802/articles/elmezeny_et_al

Escartin, M. C. P. D. (2015). Rogue cops among rogues: Trolls and trolling in social networking sites. *Philippine Sociological Review*, *63*, 169–190. https://www.jstor.org/stable/24717164

Fayard, A., & DeSanctis, G. (2005). Evolution of an online forum for knowledge management professionals: A language game analysis. *Journal of Computer-Mediated Communication*, *10*(4), JCMC1045. https://doi.org/10.1111/j.1083-6101.2005.tb00265.x

Fichman, P., & Sanfilippo, M. R. (2015). The bad boys and girls of cyberspace—how gender and context impact perception of and reaction to trolling. *Social Science Computer Review*, *33*(2), 163–180. https://doi.org/10.1177/0894439314533169

Filieri, R. (2016). What makes an online consumer review trustworthy? *Annals of Travel Research*, *58*, 46–64. https://doi.org/10.1016/j.annals.2015.12.019

Fornacciari, P., Mordonini, M., Poggi, A., Sani, L., & Tornaiuolo, M. (2018). A holistic system for troll detection on Twitter. *Computers in Human Behavior*, *89*, 258–268. https://doi.org/10.1016/j.chb.2018.08.008

Gallarza, M. G., & Gil, I. (2008). The concept of value and its dimensions: A tool for analysing tourism experiences. *Tourism Review*, *63*(3), 4–20.

Gkritzali, A., Mavragani, E., & Gritzalis, D. (2019). Negative MWOM and value co-destruction during destination crises. *Business Project Management Journal*. https://doi.org/10.1108/BPMJ-07-2019-0278

Gohary, A., Hamzelu, B., & Pourazizi, L. (2016). A little bit more value creation and a lot of less value destruction! Exploring service recovery paradox in value context: A study in travel industry. *Journal of Hospitality and Tourism Management*, *29*, 189–203. https://doi.org/10.1016/j.jhtm.2016.09.001

Golf-Papez, M., & Veer, E. (2017). Don't feed the trolling: Rethinking how online trolling is being defined and combated. *Journal of Marketing Management*, *33*(15–16), 1336–1354. https://doi.org/10.1080/0267257X.2017.1383298

Graham, E. (2019). Boundary maintenance and the origins of trolling. *New Media & Society*, *21*(9), 2029–2047. https://doi.org/10.1177/1461444819837561

Gully, A. (2012). It's only a flaming game: A case study of Arabic computer-mediated communication. *British Journal of Middle Eastern Studies*, *39*(1), 1–18. https://doi.org/10.1080/13530194.2012.659440

Hardaker, C. (2010). Trolling in asynchronous computer-mediated communication: From user discussions to academic definitions. *Journal of Politeness Research*, *6*(2), 215–242. https://doi.org/10.1515/jplr.2010.011

Hardaker, C. (2015). I refuse to respond to this obvious troll: An overview of responses to perceived trolling. *Corpora*, *10*(2), 201–229. https://doi.org/10.3366/cor.2015.0074

Herring, S., Job-Sluder, K., Scheckler, R., & Barab, S. (2002). Searching for safety online: Managing "trolling" in a feminist forum. *The Information Society*, *18*(5), 371–384. https://doi.org/10.1080/01972240290108186

Hinduja, S., & Patchin, J. W. (2013). Social influences on cyberbullying behaviors among middle and high school students. *Journal of Youth and Adolescence*, *42*(5), 711–722. https://doi.org/10.1007/s10964-012-9902-4

Hoff, D. L., & Mitchell, S. N. (2009). Cyberbullying: Causes, effects, and remedies. *Journal of Educational Administration*, *47*(5), 652–665. https://doi.org/10.1108/09578230910981107

Hong, F., & Cheng, K. (2018). Correlation between university students' online trolling behaviour and online trolling victimization forms, current conditions, and personality traits. *Telematics and Informatics*, *35*(2), 397–405. https://doi.org/10.1016/j.tele.2017.12.016

Hopkinson, C. (2013). Trolling in online discussions: From provocation to community-building. *Brno Studies in English*, *39*(1), 5–25. https://doi.org/10.5817/BSE2013-1-1

Huang, Y., & Chou, C. (2010). An analysis of multiple factors of cyberbullying among junior high school students in Taiwan. *Computers in Human Behavior*, *26*(6), 1581–1590. https://doi.org/10.1016/j.chb.2010.06.005

Hwang, Y., Jani, D., & Jeong, H. K. (2013). Analyzing international tourists' functional information needs: A comparative analysis of inquiries in an on-line travel forum. *Journal of Business Research*, *66*(6), 700–705. https://doi.org/10.1016/j.jbusres.2011.09.006

Jarvi, H., Kahkonen, A., & Torvinen, H. (2018). When value co-creation fails: Reasons that lead to value co-destruction. *Scandinavian Journal of Management*, *34*(1), 63–77. https://doi.org/10.1016/j.scaman.2018.01.002

Jarvi, H., Keranen, J., Ritala, P., & Vilko, J. (2020). Value co-destruction in hotel services: Exploring the misalignment of cognitive scripts among customers and providers. *Tourism Management*, *77*, Article 104030. https://doi.org/10.1016/j.tourman.2019.104030

Jenks, C. (2019). Talking trolls into existence: On the floor management of trolling in online forums. *Journal of Pragmatics*, *143*, 54–64. https://doi.org/10.1016/j.pragma.2019.02.006

Johnson, C. M. (2001). A survey of current research on online communities of practice. *The Internet and Higher Education*, *4*(1), 45–60. https://doi.org/10.1016/S1096-7516(01)00047-1

Jordan, P. J., & Troth, A. C. (2004). Managing emotions during team problem solving: Emotional intelligence and conflict resolution. *Human Performance*, *17*(2), 195–218. https://doi.org/10.1207/s15327043hup1702_4

Karl, M., Reintinger, C., & Schmude, J. (2015). Reject or select: Mapping destination choice. *Annals of Tourism Research*, *54*, 48–64. https://doi.org/10.1016/j.annals.2015.06.003

Kavanagh, E., Jones, I., & Sheppard, L. (2016). Towards typologies of virtual maltreatment: Sport, digital cultures & dark leisure. *Leisure Studies*, *35*(6), 783–796. https://doi.org/10.1080/02614367.2016.1216581

Kim, K. A., Byon, K. K., & Baek, W. (2019). Customer-to-customer value co-creation and co-destruction in sporting events. *The Services Industries Journal*. https://doi.org/10.1080/02642069.2019.1586887

Komppula, R. (2005). Pursuing customer value in tourism—a rural tourism case-study. *Journal of Hospitality & Tourism*, *3*(2), 83–104.

Kristensen, A. E. (2013). Travel and social media in China: From transit hubs to stardom. *Tourism Planning & Development*, *10*(2), 169–177. https://doi.org/10.1080/21568316.2013.783736

Lamberton, C., & Stephen, A. T. (2016). A thematic exploration of digital, social media, and mobile marketing: Research evolution from 2000 to 2014 and an agenda for future inquiry. *Journal of Marketing*, *80*(6), 146–172. https://doi.org/10.1509/jm.15.0415

Leung, K. (1988). Some determinants of conflict avoidance. *Journal of Cross-Cultural Psychology*, *19*(1), 125–136. https://doi.org/10.1177/0022002188019001009

Leung, X. Y., & Bai, B. (2013). How motivation, opportunity, and ability impact travelers' social media involvement and revisit intention. *Journal of Travel & Tourism Marketing*, *30*(1–2), 58–77. https://doi.org/10.1080/10548408.2013.751211

Linvill, D. L., & Warren, P. (2019). Russian trolls can be surprisingly subtle, and often fun to read. https://www.washingtonpost.com/outlook/russian-trolls-can-be-surprisingly-subtle-and-often-fun-to-read/2019/03/08/677f8ec2-413c-11e9-9361-301ffb5bd5e6_story.html?noredirect=on&utm_term=.3b893f6def6a

Lopes, B., & Yu, H. (2017). Who do you troll and why: An investigation into the relationship between the dark triad personalities and online trolling behaviours towards popular and less popular Facebook profiles. *Computers in Human Behavior*, *77*, 69–76. https://doi.org/10.1016/j.chb.2017.08.036

Loughrey, C. (2016). Tourist climbs and smashes 126-year-old statue while taking a selfie. http://www.independent.co.uk/arts-entertainment/art/news/tourist-climbs-and-smashes-126-year-old-statue-while-taking-a-selfie-a7022016.html

Lubman, D. I., Droste, N., Pennay, A., Hyder, S., & Miller, P. (2014). High rates of alcohol consumption and related harm at schoolies week: A portal study. *Australian and New Zealand Journal of Public Health*, *38*(6), 536–541. https://doi.org/10.1111/1753-6405.12266

Lumsden, K., & Morgan, H. (2017). Media framing of trolling and online abuse: Silencing strategies, symbolic violence, and victim blaming. *Feminist Media Studies*, *17*(6), 926–940.

Lund, N. F., Scarles, C., & Cohen, S. A. (2020). The brand value continuum: Countering co-destruction of destination branding in social media through storytelling. *Journal of Travel Research*, *59*(8), 1506–1521. https://doi.org/10.1177%2F0047287519887234

Luo, J. G., Wong, I. A., King, B., Liu, M. T., & Huang, G. (2019). Co-creation and co-destruction of service quality through customer to customer interactions: Why prior experience matters. *International Journal of Contemporary Hospitality Management*, *31*(3), 1309–1329. https://doi.org/10.1108/IJCHM-12-2017-0792

Malone, S., McKechnie, S., & Tynan, C. (2018). Tourists' emotions as a resource for customer value creation, cocreation, and destruction: A customer-grounded understanding. *Journal of Travel Research*, *57*(7), 843–855. https://doi.org/10.1177/0047287517720118

March, E. (2019). Psychopathy, sadism, empathy, and the motivation to cause harm: New evidence confirms malevolent nature of the internet troll. *Personality and Individual Differences*, *141*, 133–137. https://doi.org/10.1016/j.paid.2019.01.001

March, E., & Marrington, J. (2019). A qualitative analysis of internet trolling. *Cyberpsychology, Behavior, and Social Networking*, *22*(3), 192–197. https://doi.org/10.1089/cyber.2018.0210

Massanari, A. L. (2019). Come for the period comics. Stay for the cultural awareness: Reclaiming the troll identity through feminist humor on Reddit's /r/TrollXChromosomes. *Feminist Media Studies*, *19*(1), 19–37. https://doi.org/10.1080/14680777.2017.1414863

McCabe, S., Li, C., & Chen, Z. (2016). Time for a radical reappraisal of tourist decision making? Toward a new conceptual model. *Journal of Travel Research*, *55*(1), 3–15. https://doi.org/10.1177/0047287515592973

Mishna, F., Cook, C., Gadalla, T., Daciuk, J., & Solomon, S. (2010). Cyber bullying behaviors among middle and high school students. *American Journal of Orthopsychiatry*, *80*(3), 362–374. https://doi.org/10.1111/j.1939-0025.2010.01040.x

Mkono, M. (2018). Troll alert!: Provocation and harassment in tourism and hospitality social media. *Current Issues in Tourism*, *21*(7), 791–804. https://doi.org/10.1080/13683500.2015.1106447

Mkono, M., & Tribe, J. (2017). Beyond reviewing: Uncovering the multiple roles of tourism social media users. *Journal of Travel Research*, *56*(3), 287–298. https://doi.org/10.1177/0047287516636236

Munar, A. M., & Jacobsen, J. K. S. (2013). Trust and involvement in tourism social media and web-based travel information sources. *Scandinavian Journal of Hospitality and Tourism*, *13*(1), 1–19. https://doi.org/10.1080/15022250.2013.764511

Munar, A. M., & Jacobsen, J. K. S. (2014). Motivations for sharing tourism experiences through social media. *Tourism Management*, *43*, 46–54. https://doi.org/10.1016/j.tourman.2014.01.012

Nicholls, S. B., & Rice, R. E. (2017). A dual-identity model of responses to deviance in online groups: Integrating social identity theory and expectancy violations theory. *Communication Theory*, *27*(3), 243–268. https://doi.org/10.1111/comt.12113

Nisbet, D. (2004). Measuring the quantity and quality of online discussion group interaction. *Journal of ELiteracy*, *1*(2), 122–139.

O'Connell, S. (2009). Splashing out after your exams. https://www.irishtimes.com/life-and-style/travel/splashing-out-after-your-exams-1.1236776

Oriarde, A., & Robinson, P. (2019). Prosuming tourist information: Asking questions on TripAdvisor. *International Journal of Tourism Research*, *21*(1), 134–143. https://doi.org/10.1002/jtr.2247

Paavola, J., Helo, T., Jalonen, H., Sartonen, M., & Huhtinen, A. (2016). Understanding the trolling phenomenon: The automated detection of bots and cyborgs in the social media. *Journal of Information Warfare*, *15*(4), 100–111. https://www.jstor.org/stable/26487554

Paris, C. M., & Teye, V. (2010). Backpacker motivations: A travel career approach. *Journal of Hospitality Marketing & Management*, *19*(3), 244–259. https://doi.org/10.1080/19368621003591350

Patterson, L. J., Allan, A., & Cross, D. (2017). Adolescent perceptions of bystanders' responses to cyberbullying. *New Media & Society*, *19*(3), 366–383. https://doi.org/10.1177/1461444815606369

Piotrowski, C. (2012). From workplace bullying to cyberbullying: The enigma of e-harassment in modern organizations. *Organization Development Journal*, *30*(4), 44–53.

Pirolli, B. (2018). Travel information online: Navigating correspondents, consensus, and conversation. *Current Issues in Tourism*, *21*(12), 1337–1343. https://doi.org/10.1080/13683500.2016.1273883

Prior, D. D., & Marcos-Cuevas, J. (2016). Value co-destruction in interfirm relationships: The impact of actor engagement styles. *Marketing Theory*, *16*(4), 533–552. https://doi.org/10.1177/1470593116649792

Quach, S., & Thaichon, P. (2017). From connoisseur luxury to mass luxury: Value co-creation and co-destruction in the online environment. *Journal of Business Research*, *81*, 163–172. https://doi.org/10.1016/j.jbusres.2017.06.015

Rauschnabel, P. A., Kammerlander, N., & Ivens, B. S. (2016). Collaborative brand attacks in social media: Exploring the antecedents, characteristics, and consequences of a new form of brand crises. *Journal of Marketing Theory and Practice*, *24*(4), 381–410. https://doi.org/10.1080/10696679.2016.1205452

Ravescroft, N., & Gilchrist, P. (2009). Spaces of transgression: Governance, discipline and reworking the carnivalesque. *Leisure Studies*, *28*(1), 35–49. https://doi.org/10.1080/02614360802127243

Richardson, J. (1995). Avoidance as an active mode of conflict resolution. *Team Performance Management*, *1*(4), 19–25. https://doi.org/10.1108/13527599510096756

Ritchie, J. R. B. (1999). Crafting a value-driven vision for a national tourism treasure. *Tourism Management*, *20*(3), 273–282. https://doi.org/10.1016/S0261-5177(98)00123-X

Robin, M. (2013). *Lonely Planet Melbourne staff in firing line as publishing company goes digital-first*. https://www.smartcompany.com.au/finance/economy/lonely-planet-melbourne-staff-in-firing-line-as-publishing-company-goes-digital-first/

Sanchez, J., Callarisa, L., Rodriguez, R. M., & Moliner, M. A. (2006). Perceived value of the purchase of a tourism product. *Tourism Management*, *27*(3), 394–406. https://doi.org/10.1016/j.tourman.2004.11.007

Sanfilippo, M., Yang, S., & Fichman, P. (2017). Trolling here, there, and everywhere: Perceptions of trolling behaviors in context. *Journal of the Association for Information Science and Technology*, *68*(10), 2313–2327. https://doi.org/10.1002/asi.23902

Slonje, R., Smith, P. K., & Frisen, A. (2013). The nature of cyberbullying, and strategies for prevention. *Computers in Human Behavior*, *29*(1), 26–32. https://doi.org/10.1016/j.chb.2012.05.024

Smith, P. K., Mahdavi, J., Carvalho, M., Fisher, S., Russell, S., & Tippett, N. (2008). Cyberbullying: Its nature and impact in secondary school pupils. *Journal of Child Psychology and Psychiatry*, *49*(4), 376–385. https://doi.org/10.1111/j.1469-7610.2007.01846.x

Snyman, R., & Loh, J. (2015). Cyberbullying at work: The mediating role of optimism between cyber-bullying and job outcomes. *Computers in Human Behavior*, *53*, 161–168. https://doi.org/10.1016/j.chb.2015.06.050

Spencer, A., & Bean, D. (2017). Female sex tourism in Jamaica: An assessment of perceptions. *Journal of Destination Marketing & Management*, *6*(1), 13–21. https://doi.org/10.1016/j.jdmm.2016.10.002

Squires, N. (2015). Tourists posing for selfie wreck 18th-century Italian monument. http://www.telegraph.co.uk/news/worldnews/europe/italy/11581657/Tourists-posing-for-selfie-wreck-18th-century-Italian-monument.html

Starmer-Smith, C. (2009). Britons abroad: Foreign Office warning over bad behaviour. http://www.telegraph.co.uk/travel/travelnews/5549091/Britons-abroad-Foreign-Office-warning-over-bad-behaviour.html

Sthapit, E. (2018). Exploring the antecedents of value no-creation: Cruise tourists' perspectives. *Anatolia*, *29*(4), 617–619. https://doi.org/10.1080/13032917.2018.1533481

Sthapit, E. (2019). My bad for wanting to try something unique: Sources of value co-destruction in the airbnb context. *Current Issues in Tourism*, *22*(20), 2462–2465. https://doi.org/10.1080/13683500.2018.1525340

Sthapit, E., & Bjork, P. (2019). Sources of value co-destruction: Uber customer perspectives. *Tourism Review*, *74*(4), 780–794. https://doi.org/10.1108/TR-12-2018-0176

Sthapit, E., & Jimenez-Barreto, J. (2019). You never know what you will get in an airbnb: Poor communication destroys value for guests. *Current Issues in Tourism*, *22*(19), 2315–2318. https://doi.org/10.1080/13683500.2018.1475469

Stieler, M., Weismann, F., & Germelmann, C. C. (2014). Co-destruction of value by spectators: The case of silent protests. *European Sport Management Quarterly*, *14*(1), 72–86. https://doi.org/10.1080/16184742.2013.865249

Synnott, J., Coulias, A., & Ioannou, M. (2017). Online trolling: The case of Madeleine McCann. *Computers in Human Behavior*, *71*, 70–78. https://doi.org/10.1016/j.chb.2017.01.053

Tayade, P. M., Shaikh, S. S., & Deshmukh, S. N. (2017). To discover trolling patterns in social media: Troll filter. *International Journal of Scientific Research in Computer Science, Engineering and Information Technology*, *2*(5), 698–703.

Thai, N. T., & Yuksel, U. (2017). Choice overload in holiday destination choices. *International Journal of Culture, Tourism and Hospitality Research*, *11*(1), 53–66. https://doi.org/10.1108/IJCTHR-09-2015-0117

Tham, A., & Wang, M. (2017). There's a troll on the information bridge! An exploratory study of deviant online behaviour impacts on tourism cosmopolitanism. *Tourism Recreation Research*, *42*(2), 258–272. https://doi.org/10.1080/02508281.2017.1298701

Thomas-Francois, K., Joppe, M., & von Massow, M. (2018). Improving linkages through a service-oriented local farmers-hotel supply chain—an explanatory case in Grenada. *Tourism Planning & Development*, *15*(4), 398–418. https://doi.org/10.1080/21568316.2017.1338200

Tjosvold, D. (2008). The conflict-positive organization: It depends upon us. *Journal of Organizational Behavior*, *29*(1), 19–28. https://doi.org/10.1002/job.473

Uriely, N., & Belhassen, Y. (2006). Drugs and risk-taking in tourism. *Annals of Tourism Research*, *33*(2), 339–359. https://doi.org/10.1016/j.annals.2005.10.009

Waldmeir, P. (2015). *China teaches its 100 m tourists some travel etiquette*. https://www.ft.com/content/5e73443c-e9d8-11e4-a687-00144feab7de

Wu, Y. (2015). Incivility on Diaoyu island sovereignty in Tianya club. *The Journal of International Communication*, *21*(1), 109–131. https://doi.org/10.1080/13216597.2014.980296

Xie, J., Song, M., & Stringfellow, A. (1998). Interfunctional conflict, conflict resolution styles, and new product success: A four-culture comparison. *Management Science*, *44*(2 Part 2), S145–S282.

Yeh, S., Fotiadis, A., Chiang, T., Ho, J., & Huan, T. (2020). Exploring the value co-destruction model for on-line deviant behaviors of hotel customers. *Tourism Management*, *33*, 100622. https://doi.org/10.1016/j.tmp.2019.100622

Yin, J., Qian, L., & Shen, J. (2019). From value co-creation to value co-destruction? The case of dockless bike sharing in China. *Transportation Research Part D*, *71*, 169–185. https://doi.org/10.1016/j.trd.2018.12.004

The Role of Industry 4.0 Tools on Museum Attributes Identification: An Exploratory Study of Thyssen-Bornemisza National Museum (Madrid, Spain)

Alicia Orea-Giner, Carmen De-Pablos-Heredero and Trinidad Vacas-Guerrero

ABSTRACT

The objective of this paper is to examine how Industry 4.0 tools help to understand museum performance and to facilitate the development of co-creation strategies. A case study explores the museum customer experience through attribute identification and attribute value as perceived by the customer. Tourist opinion and perception form a crucial factor that can be explored through electronic word-of-mouth (eWOM). Content analysis can be used to detect attributes as perceived by tourists, with a view to improving the tourist experience. The detected attributes are evaluated in a roundtable discussion with experts and key stakeholders. The exploratory approach reported here examines the introduction of Industry 4.0 tools like eWOM to study museum visitor perception and perceived value. The results are key for designing co-creation strategies and providing precious information for implementing innovative processes such as co-creation to improve the decision-making process for museum management.

1. Introduction

Industry 4.0, also called the Fourth Industrial Revolution, has to do with the use of technology to automate production processes in industry (Xu et al., 2018b). Information and communication technology (ICT) plays an important role in Industry 4.0 (Lu, 2017) and paves the way to other technologies, such as cloud computing, the Internet of Things and social media, to facilitate production and decision-making processes based on data science results (Autio et al., 2018; Para et al., 2018). Some research projects on ICT, its development and its application in Industry 4.0 are linked to the concept of open innovation (Chesbrough, 2006; West & Bogers, 2017), which in turn has been connected to innovative management strategies (Peris-Ortiz & Hervás-Oliver, 2014) and consumers' collaborative participation in innovation (Kohler & Chesbrough, 2019) through the concept of co-creation (Von Hippel, 1986; Herstatt & Von Hippel, 1992; Prahalad & Ramaswamy, 2000; Sharma et al., 2016). Other important factors are the creation of value by consumers

(Kagermann, 2015), social media and electronic word-of-mouth (eWOM). These concepts have been analysed to reach conclusions on management processes in Industry 4.0 (Jackson, 2019; Li et al., 2018; Yoon et al., 2019).

The tourism industry has been gradually incorporating Industry 4.0 tools as well as technologies such as robotics, artificial intelligence and service automation (Ivanov & Webster, 2017; Ivanović et al., 2016). The industry is using ICT to co-create and offer services that cater better to consumer preferences in the experience society (Hanafiah & Zulkifly, 2019; Marasco et al., 2018; Peceny et al., 2019). Tourism studies point out multiple possibilities afforded by eWOM analysis techniques (Pihlaja et al., 2017). Industry researchers are applying digital technologies to the study of consumer behaviour and perceived value (Wang et al., 2017).

Another innovative technique, co-creation, influences visitor experience by providing positive emotions, thus increasing museum visitor satisfaction and loyalty (Ruiz-Alba et al., 2019). The creation of mutual value is fundamental and is linked to visitors' experience. Museums must motivate consumers to participate in co-creation activities to create value together (Thyne & Hede, 2016).

This paper applies Industry 4.0 tools and the co-creation concept to gain in-depth insight into the museum customer experience through museum attributes and their value as perceived by the customer. These tools are used to validate the proposed model of co-creation experiences in museums.

RQ1: How do Industry 4.0 methods help to understand museum attributes?

RQ2: How does attribute analysis help to promote co-creation and innovation strategies in museums?

RQ3: How do eWOM and the roundtable make it easier to identify museum attributes, leading to co-creation and innovative strategies?

The implications of the research are both theoretical, in that a method is offered to scholars and practitioners, and practical, in that the method can be applied in other contexts and museums and can help improve service quality.

This paper is divided into five sections. Section one presents the research background, a summary of theorists' contributions on typical Industry 4.0 tools and their application to co-creation in the tourism field. Section two deals with the data and methods used. Section three reports the results of the eWOM data analysis and the roundtable discussion. Section four outlines and discusses the findings. Lastly, section five gives the conclusions, practical implications and limitations of this research.

2. Literature review

The concept of Industry 4.0 calls for both workers and consumers to participate in decision-making processes (Dalenogare et al., 2018; Pilloni, 2018). Industry 4.0 tools make it possible to obtain interesting online data that can be mined and analysed for subsequent use in practical applications (Pilloni, 2018). In addition, they provide mechanisms for improving and facilitating companies' competitiveness (Adamik & Nowicki, 2018)

The concept of co-creation is implemented through the inclusion of technologies specific to Industry 4.0 (Adamik & Nowicki, 2018). Business model innovation (BMI), which relies heavily on both co-creation and the advances of Industry 4.0, considers the value of the firm itself, the value created connected to the consumer's experience and the value obtained through the experiential flow (Keiningham et al., 2019). Value co-creation enables a business to develop strategies for its survival and future growth, because value co-cocreation is a personalised response to the client's own value proposition (Sheth, 2019). Co-creation has been introduced in the tourism industry in order to develop new services and apply innovation in management (Xu et al., 2018a). To develop co-creation as an integral part of service, it is necessary to consider the interrelationship between consumers, workers, the company and technology (Fitzgerald et al., 2014; Sarmah et al., 2018).

The eWOM concept is connected to Industry 4.0, as it uses data science to analyse opinions formed from consumer's experiences (Litvin et al., 2018). Online reviews are considered a type of eWOM (Liu et al., 2018). It is generally agreed that TripAdvisor reviews are considered a source of eWOM information on consumer perception and experience (Yoon et al., 2019). Much of the debate over the value and representativeness of using eWOM revolves around the significance and the objectivity of TripAdvisor reviews as a source of information (Ayeh et al., 2013; Qiu et al., 2012; Yan et al., 2018).

EWOM analysis is connected with the term "co-creation" (Mikalef et al., 2018; Xie et al., 2019). Researchers have demonstrated the importance of co-creation in tourism (Prebensen & Foss, 2011; Sfandla & Björk, 2013). EWOM analysis has an impact on purchase intention, so it affects consumer confidence. Introducing co-creation through eWOM also affects purchase intention, increasing sales volume and revenues (See-To & Ho, 2014). There are, therefore, positive relationships between the joint creation of value for customers and the development of co-creation strategies. Client knowledge is fundamental to develop co-creation strategies, so making it easy for consumers to share their knowledge and personal experiences through online communities is essential (Frempong et al., 2020). Furthermore, negative eWOM is not the only useful kind; positive eWOM is crucial for creating links with customers and transmitting a brand's corporate image, i.e. developing relationship marketing strategies. Therefore, it is necessary to collect consumers' thoughts and feelings to obtain vital information and then foster that emotional link with the brand and co-create (Seifert & Kwon, 2019). Supply-side innovation and valuable results can be based on the construction of a co-creation platform that combines the use of innovation and technology to interact with clients (Waseem et al., 2018).

Sugathan and Ranjan (2019) stressed the critical role of customers' co-creation in tourism experiences. Buhalis and Inversini (2014) found parallels between co-creation and eWOM. Past studies have yielded some valuable insights into involving experts and the local community in roundtable discussions to spur the co-creation of tourism experiences and sustained year-round interest (Buhalis & Foerste, 2015; Trunfio & Della Lucia, 2019). In the field of tourism, online contact also encourages co-creation and the emergence of a brand-centred community, providing greater customer satisfaction and promoting brand loyalty (Rihova et al., 2018).

A considerable amount of research has focused on using eWOM studies to identify attributes of tourism destinations and resources (Xiang et al., 2017), but museum attribute perception is analysed by different methods (Burton et al., 2009; Kim & Lee, 2019; Kinghorn & Willis, 2008). Attributes influence the visitor experience, and through eWOM museums can develop more complete experiences based on visitor opinions (Zanibellato et al., 2018). To develop hypotheses about behaviour and attitudes, qualitative methods are recommended. So far, qualitative research focuses on content analysis as a method for analysing online TripAdvisor reviews (Camprubí & Coromina, 2016; Krippendorff, 2018).

The review of the literature shows a line of research focused on museum attribute identification and eWOM (Antón et al., 2019; Moran & Muzellec, 2017). The role that museums play in innovation (De-Miguel-Molina et al., 2019) is fundamental due to their connection with the local community, their promotion of cultural enterprises and their positive effect against tourism seasonalisation (Pop et al., 2019). Therefore, the concept of co-creation makes it possible to design experiences based on sustainability, increasing the value visitors perceive the different attributes of the museum as having.

Museum visitors' experience is connected to the process of co-creation. Three stages have been identified in the museum visitor experience: visit preparation, the visit and the post-visit. Co-creation has to do with the pre-visit and visit stages. The processes whereby visitors interact with museum staff during the visit are key to analysing visitors' needs and improving their experience. Furthermore, the post-visit stage is also crucial, as it facilitates greater involvement by the museum and forges a link with visitors (Antón et al., 2018). To enable co-creation, the idea is to promote use of the museum website, since the museum can reach out to tourists and visitors with maps, augmented reality, interactive activities and games on line. Moreover, during the visit it is important for the museum to encourage co-creation by using intelligent technologies and stimulating visitor participation (Buonincontri et al., 2017).

3. The context: Thyssen-Bornemisza National museum

The Thyssen-Bornemisza National Museum is located on Madrid's Paseo del Prado in an area called "the Art Walk", near the underground system's Estación del Arte (Art Station). The museum lies in a highly touristic area, one it shares with the Prado National Museum, the Reina Sofia National Art Centre (Figure 1) and other renowned museums and cultural institutions, such as Caixa Forum Madrid. The Prado-Retiro area is registered as a candidate for UNESCO World Heritage Site status.

This museum has a rich permanent collection featuring artworks by Dürer, Titian, Rubens and Cézanne. It also holds successful, high-quality temporary exhibitions. For these reasons, as a case study the Thyssen-Bornemisza National Museum can provide us with crucial information on attribute identification through eWOM and on the development of possible co-creation policies. It is a publicly owned museum whose management strategies call for it to be an open space that promotes integration, accessibility, inclusion, diversity and dialogue. Furthermore, it pursues strategies focused on both sustainability and innovation.

Besides, the Thyssen-Bornemisza National Museum ranks high in terms of visitors. In 2019 it had an 11.5% increase in visitors to a total of 1,034,941 individuals and was the third most-visited museum in Madrid after the Reina Sofia National Art Centre (1,714,049 visitors in 2019) and the Prado National Museum (3,203,417 visitors). In terms of visitor figures at the national level, it is in fifth position (Museo Nacional Centro de Arte Reina Sofía, 2020; Museo Nacional del Prado, 2020; Museo Nacional Thyssen-Bornemisza, 2020).

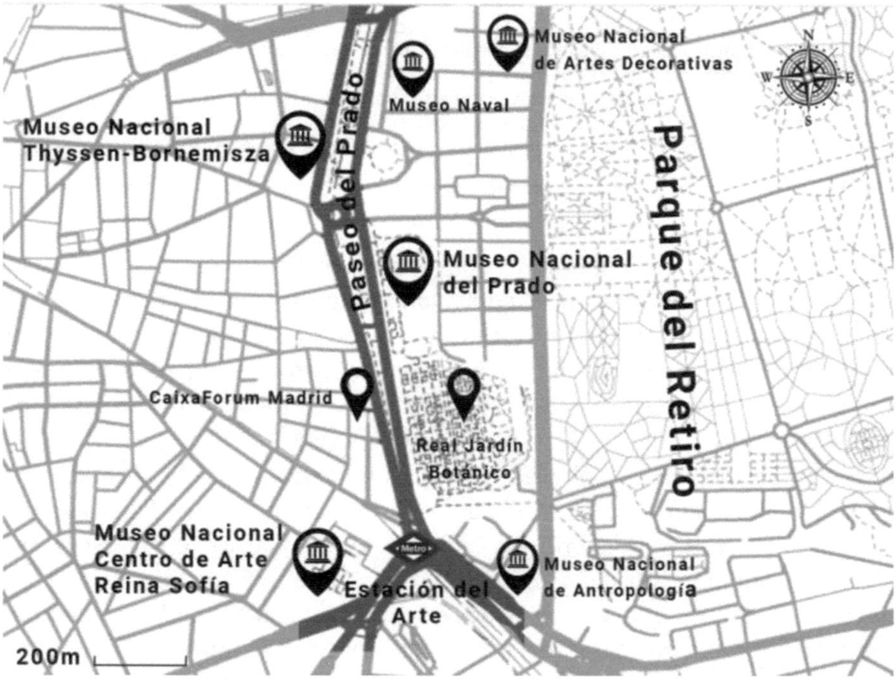

Figure 1. Plan of the area where the museum is located.

4. Methods

The method for analysing the attributes of the Thyssen-Bornemisza National Museum is derived from Industry 4.0. The procedures posited by Kinghorn and Willis (2008) and Orea-Giner et al. (2019) are modified to use eWOM for analysing museum attributes, with the addition of a roundtable discussion to confirm the identified attributes (RQ1).

The first step is to identify the most outstanding attributes cited on TripAdvisor (Kim & Lee, 2019; Zanibellato et al., 2018). The reviews are extracted via WebHarvy from TripAdvisor, the most important online travel community (Egresi & Prakash, 2019). The sample consists of 5000 reviews (2500 opinions in English and 2500 opinions in Spanish). The data are analysed automatically using Nvivo12. The attribute identification is based on the keywords (n=500) related to museum attributes (Table 1).

The second step in identifying attributes was to organise a roundtable discussion with experts in order to identify the museum's leading attributes. The participants of the roundtable discussion were museum and tourism industry professionals who were

Table 1. Text mining parameters.

Sample	Total: 5000 Spanish: 2500 English: 2500
Date	Spanish: From 24 February 2016–25 March 2019 English: From 15 July 2015–24 March 2019
Analysis	Data extracted with WebHarvy software and analysed using Nvivo 12
Keywords	500
Attributes detected	26

Table 2. Key topics of the roundtable discussion.

Topic 1	What is your opinion about the identified attributes? What strategies would you implement to improve these attributes and the museum's offer?
Topic 2	What attributes do you believe best describe the Thyssen-Bornemisza National Museum? How do you think the museum's innovation strategies affect these attributes?
Topic 3	What attributes do you think are more interesting for tourists? How can tourist perception be used in the planning of co-creation strategies?

asked to talk about the leading attributes of the Thyssen-Bornemisza National Museum. The objective was to examine how innovation strategies and co-creation affect the museum's various attributes from the point of view of the stakeholders participating in the group. Stakeholder collaboration is a fundamental part of attribute analysis and co-creation (Farr, 2016). Here stakeholder collaboration takes the form of discussion of the results of eWOM analysis.

The roundtable discussion took place at Rey Juan Carlos University's School of Social Sciences in Madrid, Spain, on 28th March 2019. The duration was one hour and 15 min. The event was video recorded, and the recordings were transcribed verbatim by the researchers. In addition, observation of the audio-visual information helped in the interpretation of the results. A homogeneous group made up the roundtable, but the participants shared a common characteristic: professional specialisation in museums and/or cultural tourism. The invitation was sent to 12 professionals, but ultimately only seven could attend. The discussion took place in the Spanish language. Participants gave written informed consent before the discussion began. The names and personal information of the participants were anonymised. The session started with a brief introduction to the topic. The participants had access to a list of the attributes selected by text mining analysis. The roundtable discussion focused on the identification of attributes, innovation and co-creation (see Table 2).

Verbal data is coded along with the attributes previously identified by text mining analysis based on frequency of repetition. The results are discussed and analysed using Nvivo 12, considering the literature on museum attributes (RQ3). The correlations between keywords are analysed (considering keywords with a result higher than 0.50) to identify links between terms. This is followed by a cluster analysis to generate diagrams based on the selected codes and cases when the clusters share words in common. This analysis is based on Pearson's correlation between keywords.

5. Results

5.1. EWOM data results

The museum attributes detected by eWOM analysis (Table 3) are based on the classification by Zanibellato et al. (2018), who divided attributes into three types core offering, peripheral services and ambience. Within each type, the different attributes perceived by tourists have been located.

Firstly, the words most mentioned with regard to the **core offering** are "collection" (61.58%), "permanent collection" and "temporary exhibitions". The attributes of the core offering are the attributes that best describe the museum, as they have 9412 mentions.

Table 3. Thyssen-Bornemisza National Museum attributes identified by eWOM analysis.

	Attributes	Word frequency	Description
1. Core offering **65.71%**	Collection	9412	The masterpieces in the museum's core offering, including the permanent collection and temporary exhibitions.
	• Permanent collection	755	The objects or pieces of art that are collected and owned by the museum.
	• Temporary exhibitions	1031	Exhibitions scheduled to open and close on specific dates.
1. Peripheral services **23.05%**	Activities	43	Activities created to promote visits to the museum and to attract, for instance, children and families.
	App	18	A smartphone app about the museum, with information on the collection and services.
	Audio guide	87	A handheld device that provides recorded information for visitors touring the museum.
	F&B* services	761	The cafeteria and restaurant inside the museum.
	Gift shop	355	A store that sells products suitable for giving as presents related to the museum.
	Guide	330	A person whose job is showing the museum or/and the book that gives visitors essential information about the museum.
	Luggage storage	5	A large locker at the museum where luggage can be left temporarily.
	Resting spaces	10	Spaces for resting during the visit.
	Staff	84	The group of people who work at the museum.
	Ticket	2147	A small piece of paper or card, usually given to show that the holder has paid to visit the museum.
	Toilets	24	A room or small building in the museum in which there are several toilets.
	Website	55	A set of pages on the Internet giving information about the museum and managed by the museum.
	Wi-Fi	9	Free Wi-Fi connection inside the museum.
1. Ambience **11.24%**	Accessibility	40	Ease with which the museum is approached, entered or used and ease with which the museum is understood.
	Building	513	The structure that contains the museum.
	Crowding	134	A large group of people visiting the museum at the same time.
	Display	35	The way the collection is arranged so it can be seen by visitors.
	Garden	44	The park and the space around the museum.
	Lighting	10	The arrangement of lights used in the museum.
	Location	306	The geographical position of the museum.
	Photos	186	The possibility of taking pictures and the procedure for taking pictures (for instance, without a flash).
	Public museum vs. private museum	399	Dichotomy between facilities provided by the government, paid for through taxes and available to everyone vs. facilities controlled or paid for a person or company and not by the government.
	Queue	248	A line of people waiting to enter the museum.

Secondly, the words used to describe the museum's **peripheral services** are "activities" (0.13%), "app" (0.11%), "audio guide" (0.57%), "F&B services" (4.97%), "Gift shop" (2.32%), "guide" (2.16%), "luggage storage" (0.03%), "staff" (0.55%), "resting spaces" (0.07%), "ticket" (14.05%), "toilets" (0.16%), "website" (0.36%) and "Wi-Fi" (0.06%).

Thirdly, the attributes referring to the **ambience** are "accessibility" (0.26%), "building" (3.35%), "crowding" (0.87%), "display" (0.21%), "garden" (0.29%), "lighting" (0.21%), "location" (2%), "photos" (1.21%), "private" (2.61%) and "queue" (1.62%). The most important correlations are related to the concept of accessibility. "Accessibility" presents a positive correlation with "collection" ($r = .36$, $p<.05$) and "luggage storage" ($r = .28$, $p<.05$). Besides, the word "storage" is correlated with "collection" ($r = .16$, $p<.05$).

5.2. Results of the roundtable discussion

Seven participants took part in the discussion (see Table 4).

These results and the classification of museum attributes in Table 3 were used as the basis for the roundtable discussion on tourist perception and co-creation possibilities. The roundtable's ideas were then analysed, considering attributes classified into three categories: core offering, peripheral services and ambience.

Regarding the core offering discussion topics, the participants assert that for tourists the collection is a fundamental element. Temporary exhibitions should be planned to take into account not only artistic criteria but also the potential for arousing interest in tourists and locals. In other words, it is possible to introduce co-creation strategies in the design of temporary exhibitions.

On the subject of peripheral services, the characteristics of the shop should be highlighted through innovative wares and designer products, and the shop should be integrated into the museum's sustainability management plan. An innovative shop that sustains visitor interest year round is fundamental in the promotion of the museum. Participants highlight the Thyssen-Bornemisza National Museum shop's capability for commercial innovation and its creation of a brand that is perceived positively due to its high quality. Participants regard the museum's communication strategies as a source of added value that enables the museum to connect with the public through social networks. Guided tours and activities are often outsourced. Even so, the staff is valued positively, and the roundtable lays emphasis on their teamwork and the museum's implementation of innovative management models.

One of the most outstanding environmental aspects is the control of seating and visits that promote sustainability and improve the visitor experience. Photographs are not allowed. The main problem with the perception of the museum and its brand image is that it is seen as being privately managed, when in fact, it is a public museum. This clash with actual fact also affects the museum's identity.

Finally, the museum engages in social action projects with different groups. The participants felt it was necessary to focus on young people, since ticket prices are high and make access to the museum difficult. There are free activities and free admission hours, but they are considered insufficient.

Table 4. Participants in the roundtable discussion.

Type	Gender	Employer or professional affiliation
Academic	Female	University. Museum studies
Curator 1	Female	National museum
Curator 2	Male	Regional museum
Museologist	Female	Spanish Association of Museology
Account manager	Male	Tour Spain
Tourism professional	Female	Madrid regional government's destination management organisation
Museum educator	Male	Private enterprise

6. Findings and discussion

The results are connected with the concept of co-creation, because the findings of this study indicate the importance of identifying museum attributes to improve visitor experience and co-creation strategies (Buhalis & Foerste, 2015; Sugathan & Ranjan, 2019; Trunfio & Della Lucia, 2019). Zanibellato et al. (2018) have identified museum attributes by studying the world's twenty leading museums in terms of visitor numbers. However, when their method is applied to a specific case study, we find that the results cannot be generalised. There are significant differences between the attributes recognised as belonging to the Thyssen-Bornemisza National Museum and the general museum attributes identified by Zanibellato et al. (2018). This is because a search focused on a single museum produces fuller, more-precise results than a general, one-size-fits-all portrait of museum attributes. These findings show that any general method needs to be applied on a museum-by-museum basis, because of museums' individual characteristics and differences (see Table 5).

The museum's collection is part of the main offer, and its connection with visitor attraction has been studied and highlighted (Dickenson, 2005; Gravari-Barbas & Fagnoni, 2015; Preziosi & Farago, 2019).

The general picture emerging from the analysis of findings is that the attributes of the Thyssen-Bornemisza National Museum identified by text mining analysis are proved and confirmed by the roundtable discussion (see Figure 2). Figure 2 takes into account the weight of each attribute and the content study, and it enables each attribute to be classified based on its relationship with aspects of co-creation.

Table 5. A comparison of attributes detected through eWOM analysis and attributes identified in previous studies.

	Attributes detected in this research	Zanibellato et al. (2018)	Burton et al. (2009)	Kinhorn & Willis (2008)
1. Core offering	Collection	✓	✓	✓
1. Peripheral services	Activities	✗	✓	✓
	App	✗	✗	✗
	Audio guide	✓	✗	✗
	F&B services	✗	✗	✗
	Gift shop	✓	✗	✗
	Guide	✓	✓	✓
	Luggage storage	✗	✗	✗
	Resting spaces	✓	✗	✗
	Staff	✓	✓	✓
	Ticket	✓	✓	✓
	Toilets	✗	✗	✗
	Website	✗	✗	✗
1. Ambience	Accessibility	✗	✗	✗
	Building	✓	✗	✗
	Crowding	✓	✗	✗
	Display	✓	✗	✗
	Garden	✗	✗	✗
	Lighting	✓	✗	✗
	Location	✗	✗	✗
	Photos	✓	✗	✗
	Public museum vs. private museum	✗	✗	✗
	Queue	✓	✗	✗

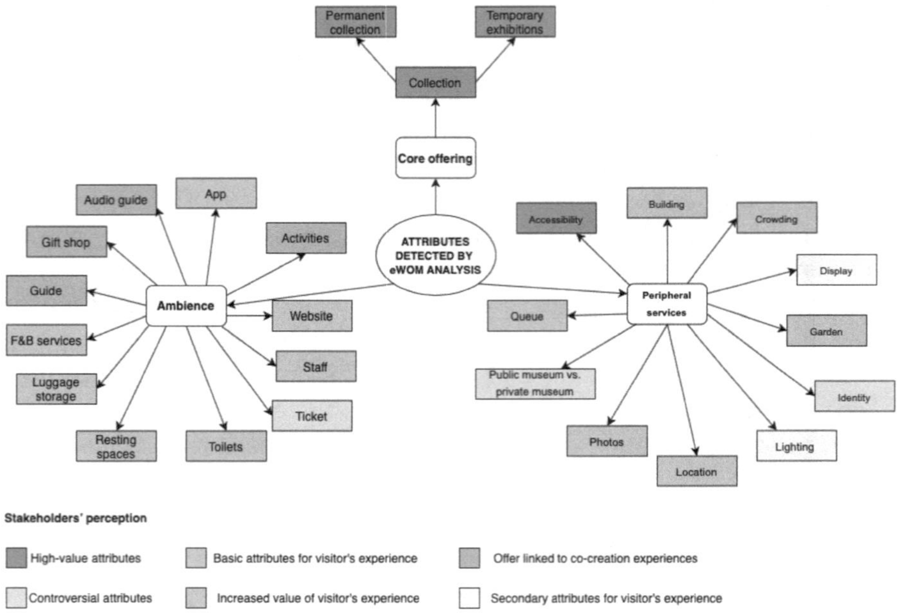

Figure 2. Attributes and co-creation experiences in museums.

An interesting side finding was that the high-value attributes of this particular museum are its collection and accessibility. The collection is the attribute that best represents the museum according to text mining analysis. Participants agreed that the "collection" attribute, including the permanent collection and temporary exhibitions, is the principal attraction for tourists. This result proves that peripheral services and ambience attributes are not undervalued. The concept of accessibility includes both the physical facilities and the ability of different segments of the market to access the museum's services (Gilmore & Rentschler, 2002) This attribute presents a crucial concept for making museums easier to understand (Haworth & Williams, 2012; Starr, 2016). The roundtable discussion did not attribute critical prevalence to this attribute.

Various attributes of peripheral services and ambience are essential to the visitor experience and thus fundamental to the offer. First, the geographical location and the building can attract visitors by offering a complete experience. Spaces play a crucial role in tourism and are a significant factor in the case of museums, which are often located at or near historically important sites and spread their image and imagination from there (Boukhris & Chapuis, 2016: Gravari-Barbas et al., 2018). Therefore, location—in this case a prime spot on the Paseo del Prado—is detected as a fundamental attribute that can attract tourists. The building is the neoclassical Villahermosa Palace, which was built in the early nineteenth century and has been converted to house the museum (Correal Avilán, 2016).

Another fundamental attribute revealed by analysis of the comments made in TripAdvisor is the museum's policy on photographs. The Thyssen-Bornemisza National Museum does not allow visitors to take pictures, so this is a point of controversy that pitches conservation considerations against freedom of visitor experience.

Museum crowding is yet a third important peripheral attribute. Kılıçarslan and Caber (2018, p. 2) remark that "Crowd perception is another determinant of overall satisfaction and behavioural intention of cultural heritage site visitors". Accordingly, museums, as premier cultural institutions, have decided to limit admission. Limited visitor admission at the Thyssen-Bornemisza National Museum has a positive significance for the roundtable participants.

Lastly, the staff is also one of the basic attributes detected by eWOM analysis. Falk and Dierking (2016) declare that the visitor experience is influenced by interactions with staff, and Antón et al. (2019) consider that staff can increase the quality of a visit and encourage visitor interaction and participation. The "staff" attribute is considered elementary, because the staff is also in charge of spreading the museum's image.

The attributes classified as secondary contributors to visitors' experience are lighting and display. Zanibellato et al. (2018) detect layout and lighting as attributes in their analysis of ten museums. In the case of the Thyssen-Bornemisza National Museum, the layout of the collection is essential. Recent changes have been made to improve lighting quality in response to visitor opinions.

The attributes classified as enabling a museum to increase the value of visitors' experience include ambience-related attributes as well. This type of attribute includes the museum's app, website, toilets, resting spaces, luggage storage and F&B services.

Lastly, there are some attributes categorised as controversial, such as ticket price, the museum's identity and the concept of public museum vs private museum. In fact, visitors' experience with a museum is not just what happens inside the museum's walls, but includes what happens on the museum's website, app and social networks (Falk & Dierking, 2016; Marty, 2007). Therefore, the museum should keep a versatile website that can be customised to address different segments. Communication through social networks is another leading factor, as well as the museum's contact with visitors before, during and after the visit. Museums introduce innovative experiences of value through their marketing and communication strategies in order to create a brand image based on design, innovation and modernity. Food and beverage services include the restaurant and the cafeteria in the museum (Mihalache, 2016). Wallace (2016, p. 238) detected that "A museum restaurant or cafeteria operates comfortably, looking at it from a branding perspective will reap many rewards: familiarisation, loyalty, repeat visits, membership, community partnerships, employee solidarity, and good standing in the scholarly universe". The food and beverage services of this museum are commended for their quality and attractiveness.

Certain attributes are identified as offering links to co-creation experiences. Museums have developed a range of activities to attract the public and create experiences that in turn encourage repeat visits (Falk & Dierking, 2016). For example, they have shops. Theobald (2000, p. 9) says, "A true museum store is a hybrid, a cross between a gift shop and a museum exhibit". The eWOM analysis and the roundtable results show that the gift shop is an outstanding attribute of the Thyssen-Bornemisza National Museum. The shop is lauded for its features and its concern for product quality. All the products for sale pursue an innovative approach that combines the participation of designers and marketing experts with demand research to create unique designer products.

Furthermore, the Thyssen-Bornemisza National Museum makes a wide variety of activities available to the local community. These activities, which are offered free of charge and are financed with public funds, are also part of the museum's offer and are a different way of approaching the museum's collection. For example, guided visits are available. Guided visits usually have two parts: "a long period of guided intensive looking, followed by a brief free period of museum cruising/exploration" (Falk & Dierking, 2016, p. 139). The quality of guided visits has been detected as an essential factor at the Thyssen-Bornemisza National Museum.

In counterpoint to the positive attributes, there are controversial attributes, such as ticket-related issues. The concept of the ticket itself is linked to ticket price, as well as free admission. EWOM analysis detects that these aspects have a negative influence on the tourist experience. Therefore, tourist opinion must be factored into a co-creation strategy that addresses these controversial attributes. Ticket price has been discussed and analysed in a number of publications (Burton et al., 2009; Ferilli et al., 2017; Frey & Steiner, 2012; Skinner et al., 2009). This point is controversial, because content analysis shows that ticket price is an element of such importance to tourists that it may decide whether they visit the museum at all. In the roundtable discussion, the stakeholders consider it necessary to understand how tourists perceive ticket price and to establish co-creation strategies to put together a complete offer adding new peripheral services or reducing ticket prices if the visitor experience is not improved.

Although it is a public museum, since the Thyssen-Bornemisza National Museum offers few hours of free admission and places restrictions on which visitors qualify for free tickets, it may easily be regarded by the public as an elitist museum. This perception has repercussions on the dissemination of the museum's image and the perception of its attributes. Therefore, tourists' perspective on this point needs to be considered. The roundtable, however, pointed out that, in addition to its established hours of free admission, the museum offers admission free of charge through complementary programmes as well. There are a number of programmes subsidised by private entities to facilitate admission to the museum. These programmes introduce innovative aspects into the visits, such as guided tours by artists.

The Thyssen-Bornemisza National Museum is a public museum, but text mining analysis reveals that it is perceived as a private museum. The roundtable highlighted this point as well and stressed its connection to the concept of identity. The concept of identity and its relationship with tourism have been studied by a great many researchers, who conclude that the identity attribute is connected with the local population's perception (Gravari-Barbas & Graburn, 2012; Nunkoo & Gursoy, 2012; Pitchford & Jafari, 2008). Gravari-Barbas et al. (2018, p. 108) say,

> The globalisation of the economy has favoured the emergence of a displaced urban elite that demands networks of cultural, educational and recreational spaces of high symbolic value: the 'brand' plays a vital role in this sense since it guarantees the symbolic value of a consumed intangible good.

In the case of the Thyssen-Bornemisza National Museum, "identity" has a strong negative link with the local population. In fact, the roundtable participants suggest the museum might engage in activities to strengthen its identity and encourage the local community to enter. To do this, broader knowledge is needed about the local population's perception of the museum and the impact the museum has, before joint co-creation strategies based on local opinions and needs can be defined.

6. Conclusion

With the application of Industry 4.0 tools, co-creation and innovation in museums can be improved. This paper reports how eWOM and stakeholder participation are useful tools for researching museum attributes to improve management's knowledge (Antón et al., 2019; Pihlaja et al., 2017). Co-creation in museums is essential for promoting the brand image and creating a unique visitor experience, which favours positive eWOM due to visitor satisfaction (Ruiz-Alba et al., 2019; Thyne & Hede, 2016).

In light of the results, some conclusions can be drawn from the eWOM results and the roundtable discussion (see Table 6). Analysing and understanding visitor experience is crucial for creating management strategies in museums (Guo et al., 2017; Keiningham et al., 2019). The participants in the roundtable discussion validated the attributes detected by data mining and discussed the possibilities of innovation and co-creation based on these attributes (RQ1). The data were analysed using a netnography approach to improve the accuracy of the results. This qualitative technique is a method that improves knowledge about tourist experiences and behaviour (Kozinets, 2019; Tavakoli & Wijesinghe, 2019).

Lastly, museums should improve their strategies for enhancing visitors' experience with peripheral services and ambience attributes, which are vital to the co-creation experience, because they complement the visitor's experience. These results suggest that the concept of co-creation could be applied to museums, especially for improving peripheral services and ambience attributes (RQ2). The stakeholder's perception, as manifested by the roundtable participants, shows that the collection and accessibility are high-value attributes. *The findings of this study suggest that museum stakeholders such as those who participated in the roundtable have valuable contributions to offer museum management* (García-Muiña et al., 2019). The following services linked to co-creation experiences can be identified: activities, audio guide, gift shop and guide, all of which are included in "ambience" (RQ3).

6.1. Practical implications

Through the study of TripAdvisor reviews, a key point has been detected, which is that tourists perceive that the Thyssen-Bornemisza National Museum is a private institution. The analysis of eWOM data and the roundtable discussion point the way to co-creation strategies that might address this misperception and other points that bear improvement.

Table 6. Research questions and answers.

RQ1: How do Industry 4.0 methods help to understand museum attributes?	They provide a better understanding of museum performance and help improve the decision-making process. The fundamental tools include eWOM analysis and co-creation through open innovation projects.
RQ2: How does attribute analysis help to promote co-creation and innovation strategies in museums?	Attribute analysis provides crucial information for identifying co-creation possibilities in the museum in order to improve visitors' experience and the value visitors perceive. Attribute analysis also points to areas ripe for the introduction of innovative strategies.
RQ3: How do eWOM and the roundtable make it easier to identify museum attributes, leading to co-creation and innovative strategies?	EWOM analysis facilitates the process of identifying museum attributes, proving the validity of Industry 4.0 tools for analysing museum attributes.

Firstly, co-creation strategies must be developed for the design of contents to interact with visitors, such as audio guides and guides. The concept of audio guides must be updated and made interactive. Guided tours should be not be limited to conventional, cookie-cutter visits, but should be based instead on visitors' concerns and needs. Therefore, personalisation is essential, as it provides added value and improves the quality of the experience. An open innovation platform could be created to consult with stakeholders and visitors about product design and activities, including the products sold in the gift shop and online. Staff, too, must be involved and motivated to participate in these initiatives and to favour intrapreneurship.

Secondly, controversial attributes must be part of this co-creation process. The concept of the ticket is linked to a unique, personalised experience. The services included in the ticket affect the visitor's experience. It is also fundamental for the museum to have a clear identity. It facilitates identification of the museum with its corporate image and its communication strategies. Both corporate image and communication must also be oriented towards the principles of co-creation.

Lastly, it is vital to consider also certain innovative attributes that are fundamental in the pre-visit and post-visit, namely, the website, the app and the F&B services. These attributes can be reconstructed based on the co-creation proposals that eventually emerge through the open innovation processes. The museum's website must be transformed from a source of static information about the museum to an integral part of the whole museum experience. In conclusion, open innovation and co-creation initiatives need to be implemented to heighten awareness of the Thyssen-Bornemisza National Museum's identity as a public museum and to combat the existing perception of the brand as private.

6.2. Limitations and future lines

The analysis was limited in several ways. First, a greater number of TripAdvisor reviews could have been collected. Second, the number of roundtable discussions could have been increased. Such changes might produce more-accurate results.

Our research is a constructive contribution toward solving the difficulty of identifying museum attributes and their value for visitors. It shows how to identify key factors with which to develop innovation and co-creation strategies in museums. This research has raised many questions in need of further examination. Future work should concentrate on analysing local community perceptions of museum attributes and the implications of museum attributes for tourists from a sustainable perspective. Therefore, by including the perception of the local community, we can get a more complete picture to use in designing strategies.

Disclosure statement

No potential conflict of interest was reported by the author(s).

ORCID

Alicia Orea-Giner http://orcid.org/0000-0001-8198-8169
Carmen De-Pablos-Heredero http://orcid.org/0000-0003-0457-3730
Trinidad Vacas-Guerrero http://orcid.org/0000-0002-7555-0985

References

Adamik, A., & Nowicki, M. (2018). Co-creating value in the Era of Industry 4.0. *Przedsiębiorczość i Zarządzanie, 19*(6, cz. 1 Emerging Challenges in Modern Management), 23–39.

Antón, C., Camarero, C., & Garrido, M. J. (2018). Exploring the experience value of museum visitors as a co-creation process. *Current Issues in Tourism, 21*(12), 1406–1425. https://doi.org/10.1080/13683500.2017.1373753

Antón, C., Camarero, C., & Garrido, M. J. (2019). What to do after visiting a museum? From post-consumption evaluation to intensification and online content generation. *Journal of Travel Research, 58*(6), 1052–1063. https://doi.org/10.1177/0047287518793040

Autio, E., Nambisan, S., Thomas, L., & Wright, M. (2018). Digital affordances, spatial affordances, and the genesis of entrepreneurial ecosystems. *Strategic Entrepreneurship Journal, 12*(1), 72–95. https://doi.org/10.1002/sej.1266

Ayeh, J. K., Au, N., & Law, R. (2013). "Do we believe in TripAdvisor?" Examining credibility perceptions and online travelers' attitude toward using user-generated content. *Journal of Travel Research, 52*(4), 437–452. https://doi.org/10.1177/0047287512475217

Boukhris, L., & Chapuis, A. (2016). Circulations, espace et pouvoir-Penser le tourisme pour penser le politique. *L'Espace Politique. Revue en Ligne de Géographie Politique et de Géopolitique*(28).

Buhalis, D., & Foerste, M. (2015). Socomo marketing for travel and tourism: Empowering co-creation of value. *Journal of Destination Marketing & Management, 4*(3), 151–161. https://doi.org/10.1016/j.jdmm.2015.04.001

Buhalis, D., & Inversini, A. (2014). Tourism branding, identity, reputation co-creation, and word-of-mouth in the age of social media. In M. M. Mariani, R. Baggio, D. Buhalis, & C. Longhi (Eds.), *Tourism management, marketing, and development* (pp. 15–40). Palgrave Macmillan. https://doi.org/10.1057/9781137354358_2

Buonincontri, P., Morvillo, A., Okumus, F., & van Niekerk, M. (2017). Managing the experience co-creation process in tourism destinations: Empirical findings from Naples. *Tourism Management, 62*, 264–277. https://doi.org/10.1016/j.tourman.2017.04.014

Burton, C., Louviere, J., & Young, L. (2009). Retaining the visitor, enhancing the experience: Identifying attributes of choice in repeat museum visitation. *International Journal of Nonprofit and Voluntary Sector Marketing, 14*(1), 21–34. https://doi.org/10.1002/nvsm.351

Camprubí, R., & Coromina, L. (2016). Content analysis in tourism research. *Tourism Management Perspectives, 18*, 134–140. https://doi.org/10.1016/j.tmp.2016.03.002

Chesbrough, H. (2006). Open innovation: A new paradigm for understanding industrial innovation. *Open Innovation: Researching a new Paradigm, 400*, 0–19.

Correal Avilán, N. (2016). *Museo Thyssen-Bornemisza: análisis crítico de la intervención y adaptación del Palacio de los Duques de Villahermosa.*

Dalenogare, L. S., Benitez, G. B., Ayala, N. F., & Frank, A. G. (2018). The expected contribution of Industry 4.0 technologies for industrial performance. *International Journal of Production Economics, 204*, 383–394. https://doi.org/10.1016/j.ijpe.2018.08.019

De-Miguel-Molina, B., Hervás-Oliver, J. L., & Boix, R. (2019). Understanding innovation in creative industries: Knowledge bases and innovation performance in art restoration organisations. *Innovation*, 1–22.

Dickenson, V. (2005). The economics of museum admission charges. In *Museum management* (pp. 116–125). Routledge.

Egresi, I., & Prakash, T. G. S. L. (2019). What makes wildlife tourists happy and what disappoints them? Learning from reviews posted on tripadvisor. *GeoJournal of Tourism and Geosites, 24*(1), 102–117. https://doi.org/10.30892/gtg.24109-346

Falk, J. H., & Dierking, L. D. (2016). *The museum experience revisited.* Routledge.

Farr, M. (2016). Co-production and value co-creation in outcome-based contracting in public services. *Public Management Review, 18*(5), 654–672. https://doi.org/10.1080/14719037.2015.1111661

Ferilli, G., Grossi, E., Sacco, P. L., & Tavano Blessi, G. (2017). Museum environments, visitors' behaviour, and well-being: Beyond the conventional wisdom. *Museum Management and Curatorship, 32*(1), 80–102. https://doi.org/10.1080/09647775.2016.1239125

Fitzgerald, M., Kruschwitz, N., Bonnet, D., & Welch, M. (2014). Embracing digital technology: A new strategic imperative. *MIT Sloan Management Review*, *55*(2), 1.

Frempong, J., Chai, J., Ampaw, E. M., Amofah, D. O., & Ansong, K. W. (2020). The relationship among customer operant resources, online value co-creation and electronic-word-of-mouth in solid waste management marketing. *Journal of Cleaner Production*, *248*, 119228. https://doi.org/10.1016/j.jclepro.2019.119228

Frey, B. S., & Steiner, L. (2012). Pay as you go: A new proposal for museum pricing. *Museum Management and Curatorship*, *27*(3), 223–235. https://doi.org/10.1080/09647775.2012.701994

García-Muiña, F. E., Fuentes-Moraleda, L., Vacas-Guerrero, T., & Rienda-Gómez, J. J. (2019). Understanding open innovation in small and medium-sized museums and exhibition halls. *International Journal of Contemporary Hospitality Management*. https://doi.org/10.1108/IJCHM-03-2018-0260

Gilmore, A., & Rentschler, R. (2002). Changes in museum management: A custodial or marketing emphasis? *Journal of Management Development*, *21*(10), 745–760. https://doi.org/10.1108/02621710210448020

Gravari-Barbas, M., Avila-Gómez, A., & Ruiz, D. C. (2018). Arquitectura, museos, turismo: La guerra de las marcas. *Revista de Arquitectura*, *20*(1), 102–114. https://doi.org/10.14718/RevArq.2018.20.1.1573

Gravari-Barbas, M., & Fagnoni, E. (2015). *Nouveaux musées, nouvelles ères urbaines, nouvelles pratiques touristiques*. Presses de l'université de Laval.

Gravari-Barbas, M., & Graburn, N. (2012). Tourist imaginaries. Via. *Tourism Review*(1). https://doi.org/10.4000/viatourism.1180

Guo, Y., Barnes, S. J., & Jia, Q. (2017). Mining meaning from online ratings and reviews: Tourist satisfaction analysis using latent dirichlet allocation. *Tourism Management*, *59*, 467–483. https://doi.org/10.1016/j.tourman.2016.09.009

Hanafiah, M., & Zulkifly, M. (2019). "Tourism destination competitiveness and tourism performance: A secondary data approach". *Competitiveness Review*, *29*(5), 592–621. https://doi.org/10.1108/CR-07-2018-0045

Haworth, A., & Williams, P. (2012). Using QR codes to aid accessibility in a museum. *Journal of Assistive Technologies*, *6*(4), 285–291. https://doi.org/10.1108/17549451211285771

Herstatt, C., & Von Hippel, E. (1992). From experience: Developing new product concepts via the lead user method: A case study in a "low-tech" field. *Journal of Product Innovation Management*, *9*(3), 213–221.

Ivanov, S. H., & Webster, C. (2017). Adoption of robots, artificial intelligence and service automation by travel, tourism and hospitality companies–a cost-benefit analysis. *Artificial Intelligence and Service Automation by Travel, Tourism and Hospitality Companies–A Cost-Benefit Analysis*.

Ivanović, S., Mijolica, V., & Roblek, V. (2016). A holistic approach to innovations in tourism. *Proceedings of ICESoS*, *2016*, 367–380.

Jackson, M. (2019). Utilizing attribution theory to develop new insights into tourism experiences. *Journal of Hospitality and Tourism Management*, *38*, 176–183. https://doi.org/10.1016/j.jhtm.2018.04.007

Kagermann, H. (2015). Change through digitization—value creation in the age of Industry 4.0. In H. Albach, H. Meffert, A. Pinkwart, & R. Reichwald (Eds.), *Management of permanent change* (pp. 23–45). Springer Gabler. https://doi.org/10.1007/978-3-658-05014-6_2

Keiningham, T., Aksoy, L., Bruce, H. L., Cadet, F., Clennell, N., Hodgkinson, I. R., & Kearney, T. (2019). Customer experience driven business model innovation. *Journal of Business Research*. https://doi.org/10.1016/j.jbusres.2019.08.003

Kim, S., & Lee, W. S. (2019). Network text analysis of medical tourism in newspapers using text mining: The South Korea case. *Tourism Management Perspectives*, *31*, 332–339. https://doi.org/10.1016/j.tmp.2019.05.010

Kinghorn, N., & Willis, K. (2008). Measuring museum visitor preferences towards opportunities for developing social capital: An application of a choice experiment to the discovery museum. *International Journal of Heritage Studies*, *14*(6), 555–572. https://doi.org/10.1080/13527250802503290

Kılıçarslan, D., & Caber, M. (2018). The impacts of perceived crowding, and atmospherics on visitor satisfaction at cultural heritage sites. *Journal of Tourism and Services*, *9*(17). https://doi.org/10.29036/jots.v9i17.25

Kohler, T., & Chesbrough, H. (2019). From collaborative community to competitive market: The quest to build a crowdsourcing platform for social innovation. *R&D Management*, *49*(3), 356–368. https://doi.org/10.1111/radm.12372

Kozinets, R. V. (2019). *Netnography: The essential guide to qualitative social media research*. SAGE Publications Limited.

Krippendorff, K. (2018). *Content analysis: An introduction to its methodology*. Sage publications.

Li, J., Xu, L., Tang, L., Wang, S., & Li, L. (2018). Big data in tourism research: A literature review. *Tourism Management*, *68*, 301–323. https://doi.org/10.1016/j.tourman.2018.03.009

Litvin, S. W., Goldsmith, R. E., & Pan, B. (2018). A retrospective view of electronic word-of-mouth in hospitality and tourism management. *International Journal of Contemporary Hospitality Management*, *30*(1), 313–325. https://doi.org/10.1108/IJCHM-08-2016-0461

Liu, X., Schuckert, M., & Law, R. (2018). Utilitarianism and knowledge growth during status seeking: Evidence from text mining of online reviews. *Tourism Management*, *66*, 38–46. https://doi.org/10.1016/j.tourman.2017.11.005

Lu, Y. (2017). Industry 4.0: A survey on technologies, applications and open research issues. *Journal of Industrial Information Integration*, *6*, 1–10. https://doi.org/10.1016/j.jii.2017.04.005

Marasco, A., De Martino, M., Magnotti, F., & Morvillo, A. (2018). Collaborative innovation in tourism and hospitality: A systematic review of the literature. *International Journal of Contemporary Hospitality Management*, *30*(6), 2364–2395. https://doi.org/10.1108/IJCHM-01-2018-0043

Marty, P. F. (2007). Museum websites and museum visitors: Before and after the museum visit. *Museum Management and Curatorship*, *22*(4), 337–360. https://doi.org/10.1080/09647770701757708

Mihalache, I. D. (2016). Critical eating: Tasting museum stories on restaurant menus. *Food, Culture & Society*, *19*(2), 317–336. https://doi.org/10.1080/15528014.2016.1178548

Mikalef, P., Pappas, I. O., Krogstie, J., & Giannakos, M. (2018). Big data analytics capabilities: A systematic literature review and research agenda. *Information Systems and e-Business Management*, *16*(3), 547–578. https://doi.org/10.1007/s10257-017-0362-y

Moran, G., & Muzellec, L. (2017). eWOM credibility on social networking sites: A framework. *Journal of Marketing Communications*, *23*(2), 149–161. https://doi.org/10.1080/13527266.2014.969756

Museo Nacional Centro de Arte Reina Sofía. (2020). *El Museo Reina Sofía aumenta un 12% los visitantes en 2019*. Retrieved May 10, 2020 from: https://www.museoreinasofia.es/prensa/nota-de-prensa/museo-reina-sofia-aumenta-12-visitantes-2019

Museo Nacional del Prado. (2020). *Datos de visitas*. Retrieved May 10, 2020 from: https://www.museodelprado.es/museo/datos-visitas

Museo Nacional Thyssen-Bornemisza. (2020). *Estrategia y resultados*. Retrieved May 10, 2020 from: https://www.museothyssen.org/transparencia/estrategia-resultados

Nunkoo, R., & Gursoy, D. (2012). Residents' support for tourism: An identity perspective. *Annals of Tourism Research*, *39*(1), 243–268. https://doi.org/10.1016/j.annals.2011.05.006

Orea-Giner, A., De-Pablos-Heredero, C., & Vacas Guerrero, T. (2019). Sustainability, economic value and socio-cultural impacts of museums: A theoretical proposition of a research method. *Museum Management and Curatorship*. 1–14. https://doi.org/10.1080/09647775.2019.1700468

Para, J., Del Ser, J., Aguirre, A., & Nebro, A. J. (2018). *Decision making in Industry 4.0 scenarios supported by imbalanced data classification*. In J. Del Ser, E. Osaba, M. Bilbao, J. Sanchez-Medina, M. Vecchio, & X. S. Yang (Eds). *Intelligent distributed computing XII. IDC 2018. Studies in computational intelligence*. Vol 798. Springer. https://doi.org/10.1007/978-3-319-99626-4_11

Peceny, U. S., Urbančič, J., Mokorel, S., Kuralt, V., & Ilijaš, T. (2019, March 1). Tourism 4.0: Challenges in marketing a paradigm shift, consumer behavior and marketing. In Matthew Reyes (Ed.), *Consumer behavior and marketing*. IntechOpen. https://doi.org/10.5772/intechopen.84762

Peris-Ortiz, M., & Hervás-Oliver, J. L. (2014). Management innovation and technological innovation: Friends or foes?. In J. L. Hervás- Oliver & M. Peris-Ortiz (Eds.), *Management innovation: Antecedents, complementarities and performance consequences* (pp. 1–17). Springer.

Pihlaja, J., Saarijärvi, H., Spence, M. T., & Yrjölä, M. (2017). From electronic WOM to social EWOM: Bridging the trust deficit. *Journal of Marketing Theory and Practice*, *25*(4), 340–356. https://doi.org/10.1080/10696679.2017.1345593

Pilloni, V. (2018). How data will transform industrial processes: Crowdsensing, crowdsourcing and big data as pillars of Industry 4.0. *Future Internet*, *10*(3), 24. https://doi.org/10.3390/fi10030024

Pitchford, S., & Jafari, J. (2008). *Identity tourism: Imaging and imagining the nation*. Emerald Group Publishing.

Pop, I. L., Borza, A., Buiga, A., Ighian, D., & Toader, R. (2019). Achieving cultural sustainability in museums: A step toward sustainable development. *Sustainability*, *11*(4), 970. https://doi.org/10.3390/su11040970

Prahalad, C. K., & Ramaswamy, V. (2000). Co-opting customer competence. *Harvard Business Review*, *78*(1), 79–90.

Prebensen, N. K., & Foss, L. (2011). Coping and co-creating in tourist experiences. *International Journal of Tourism Research*, *13*(1), 54–67. https://doi.org/10.1002/jtr.799

Preziosi, D., & Farago, C. (2019). *Grasping the world: The idea of the museum*. Routledge.

Qiu, L., Pang, J., & Lim, K. H. (2012). Effects of conflicting aggregated rating on eWOM review credibility and diagnosticity: The moderating role of review valence. *Decision Support Systems*, *54*(1), 631–643. https://doi.org/10.1016/j.dss.2012.08.020

Rihova, I., Buhalis, D., Gouthro, M. B., & Moital, M. (2018). Customer-to-customer co-creation practices in tourism: Lessons from customer-dominant logic. *Tourism Management*, *67*, 362–375. https://doi.org/10.1016/j.tourman.2018.02.010

Ruiz-Alba, J. L., Nazarian, A., Rodríguez-Molina, M. A., & Andreu, L. (2019). Museum visitors' heterogeneity and experience processing. *International Journal of Hospitality Management*, *78*, 131–141. https://doi.org/10.1016/j.ijhm.2018.12.004

Sarmah, B., Kamboj, S., & Kandampully, J. (2018). Social media and co-creative service innovation: An empirical study. *Online Information Review*, *42*(7), 1146–1179. https://doi.org/10.1108/OIR-03-2017-0079

See-To, E. W., & Ho, K. K. (2014). Value co-creation and purchase intention in social network sites: The role of electronic word-of-mouth and trust–A theoretical analysis. *Computers in Human Behavior*, *31*, 182–189. https://doi.org/10.1016/j.chb.2013.10.013

Seifert, C., & Kwon, W. S. (2019). SNS eWOM sentiment: Impacts on brand value co-creation and trust. *Marketing Intelligence & Planning*, *38*(1), 89–102. https://doi.org/10.1108/MIP-11-2018-0533

Sfandla, C., & Björk, P. (2013). Tourism experience network: Co-creation of experiences in interactive processes. *International Journal of Tourism Research*, *15*(5), 495–506. https://doi.org/10.1002/jtr.1892

Sharma, S., Conduit, J., Karpen, I. O., Hill, S. R., & Farrelly, F. (2016). Co-creation in a service innovation context. In C. Campbell, & J. Ma (Eds.), *Looking forward, looking back: Drawing on the past to shape the future of marketing*. Developments in marketing science: Proceedings of the academy of marketing science (pp. 217–217). Springer. https://doi.org/10.1007/978-3-319-24184-5_55

Sheth, J. N. (2019). Customer value propositions: Value co-creation. *Industrial Marketing Management*.

Skinner, S. J., EkelundJrR. B., & Jackson, J. D. (2009). Art museum attendance, public funding, and the business cycle. *American Journal of Economics and Sociology*, *68*(2), 491–516. https://doi.org/10.1111/j.1536-7150.2009.00631.x

Starr, R. E. (2016). *Accessibility practices & the inclusive museum: Legal compliance, professional standards, and the social responsibility of museums*.

Sugathan, P., & Ranjan, K. R. (2019). Co-creating the tourism experience. *Journal of Business Research*, *100*, 207–217. https://doi.org/10.1016/j.jbusres.2019.03.032

Tavakoli, R., & Wijesinghe, S. N. (2019). The evolution of the web and netnography in tourism: A systematic review. *Tourism Management Perspectives*, *29*, 48–55.

Theobald, M. M. (2000). *Museum store management*. AltaMira Press.

Thyne, M., & Hede, A. M. (2016). Approaches to managing co-production for the co-creation of value in a museum setting: When authenticity matters. *Journal of Marketing Management*, *32*(15-16), 1478–1493. https://doi.org/10.1080/0267257X.2016.1198824

Trunfio, M., & Della Lucia, M. (2019). Co-creating value in destination management Levering on stakeholder Engagement. *e-Review of Tourism Research*, *16*(2/3).

Von Hippel, E. (1986). Lead users: A source of novel product concepts. *Management Science*, *32*(7), 791–805.

Wallace, M. (2016). *Museum branding: How to create and maintain image, loyalty, and support*. Rowman & Littlefield.

Wang, P., Zhang, X., Suomi, R., & Sun, C. (2017). Determinants of customers' eWOM behaviour—A system Success perspective. In R. Schegg & B. Stangl (Eds.), *Information and communication technologies in tourism 2017* (pp. 401–415). Springer. https://doi.org/10.1007/978-3-319-51168-9_29

Waseem, D., Biggemann, S., & Garry, T. (2018). Value co-creation: The role of actor competence. *Industrial Marketing Management*, *70*, 5–12. https://doi.org/10.1016/j.indmarman.2017.07.005

West, J., & Bogers, M. (2017). Open innovation: Current status and research opportunities. *Innovation*, *19*(1), 43–50. https://doi.org/10.1080/14479338.2016.1258995

Xiang, Z., Du, Q., Ma, Y., & Fan, W. (2017). A comparative analysis of major online review platforms: Implications for social media analytics in hospitality and tourism. *Tourism Management*, *58*, 51–65. https://doi.org/10.1016/j.tourman.2016.10.001

Xie, X. Z., Tsai, N. C., Xu, S. Q., & Zhang, B. Y. (2019). Does customer co-creation value lead to electronic word-of-mouth? An empirical study on the short-video platform industry. *The Social Science Journal*, *56*(3), 401–416. https://doi.org/10.1016/j.soscij.2018.08.010

Xu, H., Liu, Y., & Lyu, X. (2018a). Customer value co-creation and new service evaluation: The moderating role of outcome quality. *International Journal of Contemporary Hospitality Management*, *30* (4), 2020–2036. https://doi.org/10.1108/IJCHM-08-2016-0467

Xu, L. D., Xu, E. L., & Li, L. (2018b). Industry 4.0: State of the art and future trends. *International Journal of Production Research*, *56*(8), 2941–2962. https://doi.org/10.1080/00207543.2018.1444806

Yan, Q., Wu, S., Zhou, Y., & Zhang, L. (2018). How differences in eWOM platforms impact consumers' perceptions and decision-making. *Journal of Organizational Computing and Electronic Commerce*, *28*(4), 315–333. https://doi.org/10.1080/10919392.2018.1517479

Yoon, Y., Kim, A. J., Kim, J., & Choi, J. (2019). The effects of eWOM characteristics on consumer ratings: Evidence from TripAdvisor. Com. *International Journal of Advertising*. 1–20. https://doi.org/10.1080/02650487.2018.1541391

Zanibellato, F., Rosin, U., & Casarin, F. (2018). How the attributes of a museum experience influence electronic word-of-mouth Valence: An analysis of online museum reviews. *International Journal of Arts Management*, *21*(1), 76–90.

Generating and Sustaining Value Through Guided Tour Experiences' Co-Creation at Heritage Visitor Attractions

Kamila Bezova and Iride Azara

ABSTRACT

Experience co-creation has been acknowledged as an important process to generate and sustain value. However, research in the arena of heritage visitor attractions remains limited. A qualitative cross-sectional design was used to assess UK heritage attractions providers' engagement with guided tour experiences' co-creation and the barriers faced in the adoption of this process. Findings from 11 interviews with visitor experience managers show most of the heritage attraction providers engage in processes of guided tour experience "co-production" rather than "co-creation". Barriers include limited knowledge, and "know-how" of value co-creation processes; financial, time, and human resource constraints. Importantly, findings show visitors' satisfaction with current arrangements influence the type of tour offering. This study reveals the need to further investigate heritage audiences' variations in preferences and suggests better sector integration in terms of knowledge sharing and best practice to fully explore the benefits and worth of value co-creation in this tourism sector.

1. Introduction

The relation between tourism experiences and value co-creation has been studied by researchers for some time now (Andrades & Dimanche, 2014; Azevedo, 2009; Binkhorst & Den Dekker, 2009; Campos et al., 2018; Chen, 2018; Grönroos, 2011; Grönross & Voima, 2012; Haahti, 2006; Pine & Gilmore, 1998; Prebensen et al., 2013; Tan et al., 2013; Vargo & Lusch, 2008).

Co-creation as a dynamic and interactive process of designing and delivering an experience in collaboration with the consumer with an aim to generate "value" for both consumers and tourism operators has amply been discussed in relation to its wide-ranging benefits. For example, allowing tourists to own the shaping of their experiences according to their interests often results in higher customer satisfaction (see Arnould et al., 2002; Prebensen & Foss, 2011). Additionally, co-created experiences allow tourism operators to improve service quality and customer satisfaction (Kim et al., 2012). Value in this context is defined as "value—in—experience" (Ramaswamy & Gouillart, 2010) meaning that value is a result of a co-created experience between the provider and the tourist (Carù & Cova, 2007; Lusch et al., 2008; Minkiewicz et al., 2013; Prebensen et al., 2013).

Heritage attractions are fundamental components of the UK heritage tourism sector and key to the overall UK tourism brand (CEBR, 2018; Historic England, 2018). However, research on experience co-creation as a process to generate and sustain value at heritage visitor attractions has received little attention (Ferrari, 2013; Jonasson & Scherle, 2012; Leask, 2010; Light, 2015; Mason, 2005; Potter, 2016; Puczkó, 2013). Furthermore, research on the challenges that heritage providers face in "creating value through engaging and connecting with the customers in a personal and memorable way" remains lacking (Chathoth et al., 2018; Fitzsimmons & Fitzsimmons, 2000, p. 1; Weiler & Black, 2015).

Building on the discussions presented above, and responding to the need for further research, this study investigated experience co-creation as a means to create and sustain value at heritage tourist attractions. It did so by focusing on guided tour experiences. Traditionally understood as the most popular and well-recognised mediums brokering the tourist encounter with the past (Lee, 2017; UNWTO, 2018; Weiler & Black, 2015); these experiences rely on multiple levels of interaction between the consumers and the heritage attractions providers, and as such, they arguably are optimal lenses through which to study processes of experience co-creation (Carù & Cova, 2007; Jonasson & Scherle, 2012; Scherle & Kung, 2010).

Therefore, in this context, the study investigated (1) the extent to which UK heritage providers jointly co-create experience value with the visitor; (2) and the barriers (if any) preventing UK heritage providers from investing in value co-creation. The results of this study add to the wider understanding of the value and worth of engaging with processes of experience co-creation in the heritage tourism sector and are particularly significant as to date limited knowledge exists from the perspective of the providers (Alves et al., 2016; Chathoth et al., 2018; Kim et al., 2012). Although this research specifically focuses on guided tour experiences this methodology can also be extended and applied to investigate other types of heritage visitor experiences offered at heritage attractions. Furthermore, this research was conducted in the context of UK historic properties. Conducting a similar study at other national and international heritage attractions would help assessing the worth of investing in customer-centric experiences and the challenges for incorporating them into the heritage visitor programming (Hu & Wall, 2005; Leask, 2010; Leask & Fyall, 2006).

2. Literature review

2.1. Generating and sustaining value through tourism experience co-creation

Co-creation in tourism has been defined as a development of tourism products in collaboration with the users and it is the dynamic interaction with the product or service that leads to the emergence of co-creative processes and a higher level of engagement and dialogue at every stage of the value creation process (Payne et al., 2008; Ramaswamy & Gouillart, 2010; Walls et al., 2011).

The now seminal work by Pine and Gilmore (1998) on the advent of "the experience economy" well illustrated the need for business providers to focus on experiences creation as a new source of economic value generation. As the authors pointed out

> when a person buys a service, he purchases a set of intangible activities carried out on his behalf. But when he buys an experience, he pays to spend time enjoying a series of memorable events that a company stages to engage him in a personal way. (1999, p. 2)

Pine and Gilmore (1998; 1999) and Gilmore and Pine (2000) proposed shift from a service to an experience economy highlighted businesses' need to move away from traditional marketing; invest into relationship marketing and embrace the recognition of the value of providing rewarding (and memorable) customer experiences for business differentiation and competitive advantage (Fournier, 1998; Walls et al., 2011).

Service marketing literature similarly emphasised the need to redraft business notions of value propositions to reflect (and capitalise on) changing consumer behaviour patterns. The important works of Grönroos (2008) and Vargo and Lusch (2006; 2008 and 2012) for example, whilst arguing that service provision is at the core of all modern economic exchanges, suggested value creation (for both consumers and providers of tourism) is to be found in the co-design and co-delivery and co-evaluation of engaging and interactive experiences (Kim et al., 2012) rather than simply in their co-production (Chathoth et al., 2014b; Lovelock & Young, 1979).

Service always comes with an experience (Johnston & Kong, 2011) and experience is created in service processes, in interactions between the customers and the service provider, and through processes of the servicescape (Carù & Cova, 2003 and Prahalad & Ramaswamy, 2004). Yet, as Payne et al. (2008) and Chathoth et al. (2018, p. 33) point out, value generation "does not involve a firm's own viewpoint of how the customer-firm interaction should take place". Rather, value is co-created (rather than co-produced) through use and in context, meaning the customer is always an active participant of the value creation process across all stages of the customer journey (Vargo & Lusch, 2006).

Different authors studied the importance of the emotional aspect of the tourist experience (see for example Arnould & Price, 1993; Schmitt, 1999). Tourism experiences are arguably different from other service experiences since they tend to be more hedonic and symbolic-oriented (Kwortnik, 2008). As Kim (2010) point out: experiences in tourism are of special importance. They allow tourists to fulfil their dreams and to build long-lasting memories (Larsen, 2007; Mehmetoglu & Engen, 2011) through engaging in activities and connecting with other people (Binkhorst & Den Dekker, 2009; Rihova et al., 2014). Tourists travel mostly voluntarily, not because they have to, but because they want to and, in so doing, they participate in the production of their own experience through their time, effort, and money (Bjőrk, 2014; Chen & Chen, 2010; Prebensen et al., 2013). As such, it is arguably not surprising that the notions and benefits of experience co-creation as a process to generate value have featured highly in recent tourism research literature (see for example Binkhorst & Den Dekker, 2009; Campos et al., 2016; Haahti, 2006; MacLeod et al., 2009; Prebensen et al., 2013; Richards & Wilson, 2006).

Tourist experiences (here arguably heritage experiences) emerge in relation to a tourist journey, as a consequence of travelling, in sequences of events (Zouni & Kouremenos, 2008). In this context, tourists are not simply passive recipients of services, rather crucial *operant resources* in the value creation process, co-applying their knowledge, skills, and expertise through their actions, processes, and performances (Alves et al., 2016; Carù & Cova, 2007; Prebensen et al., 2013).

The underpinning benefit of co-creating tourism experiences is that it enables "the organization and the customer to use each other as a productive resource to co-shape the expectations of the latter" (Chathoth et al., 2014b, p. 34). In this sense, co-created tourism experiences- firmly centred in the interaction of people, products, and services alter experience value for both visitors and organisations and as such they position themselves as crucial mechanisms to not simply generate, rather sustain long term business value in tourism (Prahalad & Ramaswamy, 2004; Prebensen et al., 2013).

2.2. The challenges to generate value through experience co-creation at heritage visitor attractions

Vargo and Lusch (2004; 2006) highlighted how organisations that fail to understand the differences between value co-production (i.e. where the organisation viewpoint still controls customers-businesses interactions; length and depth of dialogues) and value co-creation may be still operating under "goods-dominant logic" and production-centric business modalities. On a similar vein Chathoth et al. (2018) argued: "to fully capitalise on value co-creation, the focus should be on anticipated engagement of customers [...] at every stage of the experience value creation process" through ongoing dialogue and for the purpose of innovation; rather than "simply a process of adding value at each stage and then marking up the price at the end of then value chain" (p. 32). Furthermore, these authors recommended transitioning towards "service-dominant logic" business modalities and co-creative service offerings requires an honest and careful analysis (on the part of the providers) of the business value proposition as well as an identification of the internal (e.g. front-line training and capacity building; technological limitations; time and costs; resistance to change, etc.) and external barriers to change (e.g. consumer's willingness and social, intellectual abilities; perceptions and attitudes towards the experience provider).

Heritage visitor attractions are a well-established and popular component of the heritage tourism sector and an important source of local and national economic growth in the UK (Boyd, 2002; UNWTO, 2018; VisitBritain, 2017). As Timothy (2011) highlights people by the hundreds of millions travel worldwide each year to seek out and experience sites of historical significance and this trend does not seem to wane away. Despite this notion, research on issues related to heritage attractions remains fragmented, mostly focusing on capturing the visitor's viewpoint rather than the management perspective (Leask, 2010; Milman, 2001; Richards, 2002). When available, research shows how processes of marketisation and "tourismification" (Salazar, 2009) pose great strains to heritage attractions providers as indeed they demand the design and delivery of heritage tourism experiences capable of satisfying new consumers sensibilities *and* protecting the interests of a wide variety of stakeholders (e.g. volunteers; public sector organisations; curators; archaeologists; local communities as well as visitors) and the heritage assets alike (du Cros & McKercher, 2020; Hewison, 1987; Jansen-Verbeke & Russo, 2008; Leask, 2010; Melpignano & Azara, 2018; Smith, 2016; Smith & Richards, 2013; Timothy, 2011).

Furthermore, it shows that the unique mix of individual resources and sites, varying levels of management and staff skills; technological infrastructure capabilities; workforces configuration (e.g. contingent; seasonal; transient) and often fluctuating patterns of visitor demand (Andrades & Dimanche, 2014; Buhalis et al., 2006; Leask, 2010, p. 159;

Richards, 2018; Timothy, 2011) adds an additional layer of complexity in the co-design and co-delivery of heritage tourism experiences (Chathoth et al., 2014b; Chen et al., 2014; Grönroos, 2008; Hansen & Mossberg, 2017; Larsen & Meged, 2013; Prebensen et al., 2013; Walls et al., 2011; Weiler & Black, 2015; Zatori, 2013). These challenges are further compounded by the systemic lack of national sector funding and limited cross-sectorial network collaboration, which arguably affect the level innovation and quality of resource deployment necessary to the effective use of the heritage assets (Costa & Buhalis, 2005; Gravari-Barbas et al., 2016; Leask, 2010; Leask & Fyall, 2006; Richards, 2002 and 2007; UNWTO, 2018).

In this operating environment, embracing the tenets of co-creation (as well as developing the technical *know-how* required) could potentially benefit the management of day-to-day activities at heritage tourism sites, increasing, for example, the overall visitors' perception of quality and satisfaction. Importantly, it could also aid in the strategic achievement of sustainable heritage tourism management goals (du Cros & McKercher, 2020; Prahalad & Ramaswamy, 2004; Smith & Richards, 2013).

However, research on experience co-creation as a process to generate and importantly to sustain value at UK heritage visitor attractions has thus far received little attention (Ferrari, 2013; Jonasson & Scherle, 2012; Light, 2015; Mason, 2005; Potter, 2016; Puczkó, 2013). Importantly, research on the challenges that UK heritage attractions providers face in creating value through engaging with experiences that connect with the customers "in a personal and memorable way" remains little investigated (Chathoth et al., 2018; Fitzsimmons & Fitzsimmons, 2000, p. 1; Leask, 2010).

Guided tours are considered a key component of the overall visitor experience at UK heritage attractions (Visit Britain, 2017) for their ability to broker visitors' encounters with the setting and other visitors and to performatively co-create places and spaces (Azara & Crouch, 2006; Jonasson & Scherle, 2012; Timothy, 2011).

As Weiler and Black (2015) point out, despite developments in information communication systems together with innovations in the arenas of augmented and immersive reality (Chu et al., 2012; Chung et al., 2018; Jung et al., 2018) heritage providers still predominantly rely on guided tours to provide enriching, educational experiences (Zatori, 2013). On a similar vein, Lee (2017) and du Cros and McKercher (2020) point out providers' mixed response to the incorporation of technology into their menu offering may also be linked to visitors' needs and characteristics. Indeed, these authors suggest, much of the bulk of heritage visitors still seek and expect to get *more traditional forms of* guided tours.

Yeoman (2012) identifies traditional forms of guided tours as (1) "group tours" predominantly targeting the "unexperienced traveller" as their role is more passive, and they perceive the guide as a presenter or entertainer who is in control. (2) experience-focused group tours which consider visitors as actors and aim to involve them either passively or actively in the experience. However, the guide is still in control of the tour experiences. And finally, (3) "personalised tours" which offer opportunities for visitors to customise their experience with the attraction provided and which also have the potential to foster processes of co-creation (i.e. both tour guides and visitors share control of the experience). Especially in this latter, the role and performance of the tour guide is central to value co-creation as visitors seek engagement in the planning, design, and production of their own experiences (Binkhorst & Den Dekker, 2009; Richards & Wilson, 2006).

Along these reasonings and because of their potential to greatly enhance visitor satisfaction and service quality perception (Carù & Cova, 2007; Prebensen et al., 2013; Scherle & Kung, 2010; Weiler & Black, 2015), guided tours can be considered as optimal lenses through which explore the issues under investigation.

3. Methods

A qualitative cross-sectional research design was deployed in order to investigate how heritage providers jointly co-create heritage experiences such as guided tours and identify the barriers (if any) preventing UK heritage attractions from investing in value co-creation. This approach was considered the most fitting to explore an area to date little investigated (Silverman, 2016). Furthermore, a cross-case analysis allowed for deeper and meaningful comparison and evaluation of the problem under investigation (Baxter & Jack, 2008; Bryman, 2012).

3.1. Participants

A criterion sampling strategy was used to identify participants to the study (Patton, 1990). Specifically, visitor experience managers of 50 UK heritage attractions were initially contacted on the basis that: (1) these attractions featured in the list of most popular UK paid historic properties measured by Visit Britain (2017) in terms of annual visitor numbers and (2) they offered a range of traditional tour guiding experiences (Yeoman, 2012). Participants were initially contacted via email requesting either a face to face or phone interview. The email advised that, should they wish to participate in the research, they could contact the researcher via email to arrange a feasible date, time for the interview. A total of 11 self-selected respondents contacted the researcher agreeing to participate in the interview (please see Table 1 for heritage providers characteristics).

3.2. Survey instrument

Semi-structured telephone interviews lasting approximately 1½ hour were conducted over a period of three months between January—March 2019. Interview questions were derived from the literature and specifically asked participants to elaborate on: (1) the type of guided tours offered and the type of visitors that used them (Weiler & Black, 2015; Yeoman, 2012) (2) to elicit heritage providers' views on value co-creation (Arnould et al., 2002; Bartella, 2014; Hansen & Mossberg, 2017; McIntyre, 2010; Minkiewicz et al., 2013; Prahalad & Ramaswamy, 2003; Prebensen et al., 2013; Prebensen & Foss, 2011). (3) to identify the barriers to generate value co-creation through the staging of guided tour experiences (Chathoth et al., 2014b; Prahalad & Ramaswamy, 2004; Prebensen et al., 2013; Weiler & Black, 2015).

3.3. Mode of analysis

Thematic analysis (Braun & Clarke, 2012; Patton, 2001) was used for systematically organising the data set; identifying commonalities of responses to the way a topic was discussed and making sense of those commonalities. In this sense the researchers followed a multi-phase iterative process which involved data familiarisation; generation

Table 1. Heritage attractions sample characteristics.

Heritage attraction	HA1	HA2	HA3	HA4	HA5	HA6	HA7	HA8	HA9	HA10	HA11
Type of heritage attraction	Museum/ historic property	Historic House/ House and Garden/Palace	Historic properties	Castle	Industrial historic property	Historic House /House and Garden	Historic House /House and Garden	Historic House /House and Garden	Historic property	Historic House/ House and Garden	Historic House/ House and Garden
Location	Somerset	Greater London	Derbyshire	Kent	Derbyshire	Shropshire	Derbyshire	Derbyshire	Yorkshire	Cheshire	Warwickshire
Numbers of visitors per year (2017)	1,300,000+	900,000+	200,000+	300,000+	200,000+	400,000+	300,000+	600,000+	400,000+	200,000+	400,000+
Type of Guided Tours (GT) offered	Public GT; private GT; Behind the scenes GT; Audio Guides; Children's GT; British sign language audio guide	Audio Tours; live interpretations and costumed GT; GT for disabled; bespoke private GT; roof GT	Costumed GT; Guided walks	Adult (public) GT; schools GT; Behind the scenes GT; Themed GT with workshop attached	Public GT; Village GT; Costumed GT; Specialist GT	Public GT; Behind the scenes GT; Re-enactment GT	Interactive/ behind the scenes GT; Guided walks	House and garden GT; Buggy tours of the gardens; Educational GT; Closed season GT; Estate walking tours; Multi-media Guide	Public GT; Garden GT; Group GT; Schools GT	Themed GT; Behind the scenes GT; Public GT; Guided walks	Public GT

of coding and finally the development of key themes (Braun & Clarke, 2006). These processes were conducted by both researchers independently of each other. The key themes emerging from the interviews were eventually cross-compared and analysed with an aim of generating the final themes. These are discussed in the section below.

4. Findings

4.1. Producing and consuming heritage guided tour experiences

Findings show heritage providers consider guided tours a key component of the heritage attraction offering as providing conveniently packaged, time-bound, educational experiences which are well recognised by a wide range of tourism audiences (Jonasson & Scherle, 2012; Weiler & Black, 2015; Zátori, 2016; Zillinger et al., 2012):

> … if you go to any attraction you see "guided tour", you know what that is, so I think a lot of people would think: I want to learn about this place, [a] guided tour would tell me everything I need to know. (HA1)

Most providers recognise that audiences are changing with some visitors seeking more participatory and entertaining experiences and that there is scope for designing new, innovative tours to generate consumer experience value across a wide range of demographics: "[…] originally [our tour] it was aimed at purely millennials but last year we found that that sort of millennial mindset […] people wanting high-quality experiences, ones that is something, maybe a VIP or tailored went across the ages" (HA7).

However, all providers acknowledge it is the more traditional forms of guided tours such as daily tours (often included in the admission price), and pre-booked (often private) tours for groups or individuals that remain the most popular types of tours demanded by heritage visitors (du Cros & McKercher, 2020). The former is identified as a "more passive experience" (Mehmetoglu & Engen, 2011; Pine & Gilmore, 1998) aimed at brokering the tourist encounter with the tangible and intangible heritage. The latter offering some opportunities for visitor engagement in the co-production and often customisation of tailor-made components of the experience (Yeoman, 2012).

Along these reasoning, all providers concede, it is the audience needs, characteristics, size, and composition rather than the demographics that heavily influence the type of service provision and the type of value creation modalities:

> I think is not about demographics, is more to do with the size of the group and how interactive it wants to be (HA2);

> Many of these visitors prefer to just listen to a talk, which is also fine (HA11);

> And lot of people come here and they don't actually know what they have come for. Most are first-time visitors, Some are on a coach trip -[and] They might not even look at our website […], so that's probably the biggest barrier (HA1);

> If people want a private tour, they can say what they are interested in. [For example] they might be a group of architects that are coming around and they'll want to know more about the design and the building and we provide a tour around this … (HA1)

Most providers highlight how visitors to heritage attractions largely arrive at the site without having made any prior contact with the provider, especially during peak season. As such the marketing strategies adopted by heritage providers reflect the nature of the demand for guided tour experiences: "we profile our visitors to a certain extent. We know that at specific times of the year or on specific days we will get predominantly one type of visitor: the one who will just turn up at the site" (HA6).

Furthermore, despite admitting that social media interaction, as well as online presence (combined with the provision of online booking systems) is important for engaging with heritage visitors; many providers acknowledge that most of the visitors still prefer to communicate and book experiences via telephone or email (Lee, 2017). This is especially so in the context of first-time visitors or those looking at gaining information and to discuss elements of tour customisation.

> We have webpage, we have very active social media, we have Instagram, and then we also go to exhibitions and things … but a lot of it [discussing tour specifics and booking] is just over the phone, the old fashion way, which is how a lot of people prefer it. (HA4)

In this respect, on-site advertising of daily guided tours is considered by providers as essential as most of the people still arrive on the site with no prior knowledge:

> … we have a welcome leaflet when they arrive at the site that tells them the kind of things they can look out for. We also tell visitors through signage that there are guided tours and the staff also tells visitors about the guided tours and what time they are (HA1);

> … every day we provide a list of different tours available on that day at visitor reception … (HA6)

Audience characteristics, needs and behaviour are also acknowledged as heavily contributing (and constraining) the regulation of and the level of interaction between tour guide and visitors during the actual service provision stage. Indeed, many providers state that most of the interaction between the guides and the visitors happen in the form of asking and answering questions. This is recognised as necessary to minimise operational time-constraints and, importantly, to satisfy the wide spectrum of audiences' needs.

> Sometimes you've got to be careful that you don't actively allow too much. Because if you do, you then start running behind (HA3);

> … asking questions are a difficult one, because you will often have people who want to ask a lot of questions so they'll be a couple of people in any group that will want to ask a lot of the questions but you have to be careful because not everyone else in the group is paying to hear someone from their group constantly talking. (HA4)

Here, the role of the tour guide is considered by all providers key in the service encounter as his/her performance directly affects visitors' experiences. Indeed, this category of tourism workers is envisioned as having the ultimate responsibility to perform the tourist gaze and amend the encounter according to visitors' willingness to interact (Urry & Larsen, 2011):

> It is part of the guide's role to read the audience and determine how much interaction the audience wants (HA11);

> If you've got a group of 50 you would not try to do that because it takes too long so you might just mention that you would have bowed 3 times to the throne rather than actually get people to do it … (HA2)

Finally, tour guides are considered vital resources to measure the quality of the service provision with all providers acknowledging that visitors are happier to give verbal feedback straight to their guide rather than using comment cards, visitor surveys or social media:

> [...] we encourage to give feedback anonymously through a comment card, email, or social media review [...] but most of the feedback are given directly to the guide (HA10);

> I say we do review our TripAdvisor and we respond to people and we then thank the guides [...] but feedback is mostly given verbally on site. (HA5)

Against this backdrop, findings suggest that most heritage providers engage with processes of guided tour experience *co-production* and individual/ group *customisation* rather than co-creation (Chathoth et al., 2018). They highlight how, from a provider's point of view the perceived visitors' satisfaction with current arrangements and their unwillingness to engage in the co-creation of guided tours is a key factor determining their lack of engagement with these processes:

> I think there might be a barrier from people who are getting in contact, who are actually not interested in co-created anything. (HA3)

> Sometimes, I think those kinds of people just aren't in the market for heritage guided tours. (HA6)

> if you want them to engage with it and pay for it, it needs to be something that they actually want. Otherwise [,,,] they would not be interested. (HA5)

4.2. Managerial and operational barriers to "leaving the beaten path"

Findings however also show how shifting towards new styles of guided tours (i.e. co-created tours) and business value modalities also presents important internal managerial and operational challenges, which further prevent providers from leaving "the beaten path" (Chathoth et al., 2018). 6 out of 11 participants for example admit they are unfamiliar with the concept of co-creation, despite appreciating the potential beneficial impacts that this process could have in terms of increasing customer satisfaction, encouraging repeat visits and even engaging traditionally disengaged groups (Melpignano & Azara, 2018).

> Interesting ... I am not familiar with it, but [with these types of tours] you may get an honest view of what visitors want to see and hear and obviously that would make it a better visit for them and that makes it more likely to return and spend more money and tell their friends. (HA5)

> It could help to encourage disengaged communities/ visitors. (HA8)

Furthermore, findings show how among those familiar with the concept and its benefits, only a few display some level of proactiveness towards cross-sectorial collaboration and information sharing:

> I think it is important to engage in sharing information and know-how on how to provide what your audience actually wants, especially at the moment. We try to be up-to-date with this consumer trend (i.e. experience co-creation). (HA7)

All of the participants identify both budgetary and time constraints as important barriers further inhibiting their ability to build *deeper* and *trustworthy* relationships with customers (Chathoth et al., 2018).

> The biggest thing is time, because co-creation and co-created tour do take time to design the processes to actually do, to get people together, I think that's a big thing and if people are willing to weigh the benefits of it (HA7);

> Time and staffing—to get a successful two-way dialogue with visitors takes a lot of time and is labour intensive. (HA11)

In this context, challenges to upskill a contingent and often heterogeneous workforce (often a mixture of FT/PT; paid staff and volunteers) are also identified as important barriers by many providers.

> I also think a big barrier, or one that needs managing properly is integration of the concept into the property programme and across the workforce that we have ... You really need to think about that and do It well and think about who will it affect, how will it affect and not just think about the visitors [... you need to] think about the volunteers and the staff and how they would respond to something that's different ... people are always worried about them becoming obsolete ... (HA7);

> It's difficult ... some guides are somewhat reluctant to adapt their style of a tour and would resist the idea of tailoring their tour to visitors, any more than they already do. (HA9)

5. Discussion

This research aimed to explore the challenges to achieve and sustain value co-creation in the context of guided tours experiences offered at heritage visitor attractions (Bartella, 2014; Chen, 2018; Minkiewicz et al., 2013). Guided tours are key, popular experiences commonly offered at heritage visitor attractions, and as such, they arguably are optimal lenses through which to begin unpacking these issues (Hansen & Mossberg, 2017; Lee, 2017; Weiler & Black, 2015).

Literature shows that service providers can use the staging of experiences as a tool for creating value through engaging with the customer in a personal way (Fitzsimmons & Fitzsimmons, 2000). Furthermore, it highlights how organisations that facilitate the active participation of customers in value co-creation processes are able to reduce investment levels (Lusch & Vargo, 2009; Vargo et al., 2008). However, Chathoth et al. (2014a) and Prahalad and Ramaswamy (2004) highlight, to fully benefit from this new approach to value creation organisations and consumers should display long-term commitment to a joint production process which is equally *owned* by both parts. That is, only when organisations and consumers use each other's operant resources (for example integrating and applying specialist competencies such as, for example, skills and knowledge to co-shape each other's expectations) that co-creation "becomes the foundation stone for business competitive advantage" (Alves et al., 2016, p. 70; Gummesson & Mele, 2010; Vargo & Lusch, 2008).

In the context of this study, however, findings reveal that heritage providers' engagement with processes of experience value co-creation remain limited (Arnould et al., 2002; Prebensen & Foss, 2011). Heritage providers are mostly still involved in processes of co-production with consumers with mainly elements of customisation being currently incorporated in their range of tour guide offering and throughout every touch point of the customer journey (Chathoth et al., 2018).

Despite findings showing an increased recognition that audiences' sensibilities and mindsets may be changing (Bartella, 2014; Prahalad & Ramaswamy, 2004; Prebensen & Foss, 2011); all the heritage providers highlight how the bulk of heritage visitors simply demand entertaining, educational experiences in the form of daily tours or pre-booked tours for individuals or groups, potentially reflecting the *massification* of the cultural heritage tourism demand (du Cros & McKercher, 2020; UNWTO, 2018). In this context, consumers' needs and characteristics are perceived as playing a key role in influencing the way organisations are deploying their operant resources (i.e. skills and competencies) to generate experience value (Alves et al., 2016; Prebensen et al., 2013; Visit Britain, 2016). Furthermore, it is the perceived visitors' satisfaction with current experience offerings as well the perceived unwillingness to engage with other value creation modalities that are identified as shaping the extent of and the level of providers' efforts to produce guided-tour experiences.

However, Alves et al. (2016, p. 71) and Payne et al. (2008) point out how: "organisations require the appropriate strategic positioning and business guidelines [...] to be able to leverage internal knowledge, skills and resources for the development of processes and practices which integrate customers' resources in the co-creation value process". Thus, the organisation' s resources make-up influences the way consumers in turn engage with and choose to use their resources (Arnould et al., 2006). In this context, limited knowledge and understanding as well as *know-how* can become important barriers to the deployment and implementation of resources (i.e. financial, human, and technological) in the most efficient and effective ways (Chathoth et al., 2014a; 2018).

Findings from this study show most heritage providers are not familiar with the concept, despite many recognising the benefits that co-created guided tours could bring in terms of increased customer engagement and satisfaction and perhaps repeat visit (Gilmore & Pine, 2000; Mathisen, 2013; Prebensen & Foss, 2011). Furthermore, they reveal how financial; time and human resource constraints are perceived by providers' as further hindering processes of change (Chen et al., 2014; Hansen & Mossberg, 2017; Larsen & Meged, 2013; Prebensen et al., 2013; Walls et al., 2011; Weiler & Black, 2015; Zatori, 2013).

Finally, findings suggest how providers' resistance to embrace step change may be further compounded by the characteristic limited cross-sector collaboration and knowledge sharing as indeed only few participants consider this mechanism important for enhancing the effective use of the heritage assets and show some level of proactiveness towards it. Whilst this is not atypical across the sector (Costa & Buhalis, 2005; Gravari-Barbas et al., 2016; Leask, 2010; Leask & Fyall, 2006; Richards, 2002 and 2007); research shows that engaging with such processes may be beneficial to foster innovation and the capturing of different market needs (UNWTO, 2018).

6. Conclusion

Value co-creation has been discussed in the academic literature and the arena of tourism for many years now (see, for example, Andrades & Dimanche, 2014; Binkhorst & Den Dekker, 2009; Dong & Siu, 2013; Chathoth et al., 2014b; Campos et al., 2016; Haahti,

2006; MacLeod et al., 2009; Mehmetoglu & Engen, 2011; Prahalad & Ramaswamy, 2004; Richard and Wilson, 2006; Prebensen et al., 2013). However, limited research exists in the context of heritage visitor attractions. Where present, research mostly focuses on understanding visitors' needs and behaviour rather than the management perspective (Leask, 2010; Milman, 2001; Richards, 2002).

Building on the above, this study set out to explore guided tour experience co-creation as a means to create and sustain value at heritage visitor attractions (Bartella, 2014; Chen, 2018; Minkiewicz et al., 2013) by conducting in-depth interviews with UK heritage attraction providers. Specifically, it investigated: (1) the extent to which UK heritage providers jointly co-create experience value with the visitor (2) and the barriers (if any) preventing UK heritage providers from investing in value co-creation.

Confirming and extending the work of Chathoth et al. (2014b; 2018), Leask (2010), and Vargo and Lusch (2004; 2006; 2012), this study shows that in the context of heritage visitor attraction, several external and internal challenges are preventing providers from engaging with these processes. These range from visitors' current level of satisfaction with current tour guides' modalities and overall unwillingness to engage in processes other than co-production and customisation; to limited providers' knowledge and understanding of co-creation; and site-specific financial, time, and human resources constraints. Furthermore, confirming and extending the works of Gravari-Barbas et al. (2016), Leask (2010); Melpignano and Azara (2018), and UNWTO (2018) this study shows that across heritage attractions providers' reluctance to engage with cross-sector collaboration and knowledge sharing is further stifling processes of innovation and change.

6.1. Theoretical implications

This study has academic implications. Firstly, it extends the knowledge and understanding of co-creation as a process to generate and sustain value at heritage visitor attractions. Heritage tourism research literature acknowledge the criticality of designing and delivering innovative heritage experiences capable of satisfying new consumers' sensibilities (Buhalis et al., 2006; Chen & Chen, 2010; Costa & Buhalis, 2005; Gravari-Barbas et al., 2016; Melpignano & Azara, 2018; Richards, 2002; Smith & Richards, 2013; Zatori, 2013). However less is known about the role co-creation can play in the delivery of these experiences (Antón et al., 2018; Prebensen & Xie, 2017). This study shows that co-creation as a process to generate business value is not fully understood and implemented across the heritage attraction sector provision. Secondly, this study contributes to a better understanding of the challenges heritage providers face in engaging with experience co-creation (Chathoth et al., 2018; Leask, 2010; Richards, 2002; UNWTO, 2018). Evidence from this study suggests their limited knowledge and understanding of experience co-creation as a mechanism to generate and sustain value at heritage attractions and further operational constraints are to a degree stifling processes of change. This study reveals the challenges faced by these providers and highlights an area of research thus far arguably overlooked.

6.2. Managerial implications

This research also has managerial implications. This study suggests providers should focus on gaining a better understanding of their visitors' needs, characteristics, and preferences as this will help to schedule their tours accordingly. Furthermore, providers should seek to gain a better understanding of the ways customers resources influence their value co-creation (Alves et al., 2016; Prebensen et al., 2013; Prebensen & Xie, 2017). In so doing they will be in better position to evaluate their role in developing these resources as well as the challenges of integrating co-creative processes effectively and efficiently within their heritage attraction programming. Heritage providers should also proactively engage with cross-sector collaboration and knowledge sharing as this would enable to share best practice on guided tours co-creation across the sector and would also aid in the recognition of where site-specific investment is needed in terms of skills; competency and overall capacity building (UNWTO, 2018).

6.3. Limitations and future research

Drawing conclusions on the findings, the process of value co-creation in the heritage tourism academic literature should be further explored from both the providers and consumers' perspective (Chathoth et al., 2018; Leask, 2010; UNWTO, 2018). As suggested by the findings, the limited UK heritage providers' knowledge and understanding of the concept together with financial, time, and human resource constraints appear to be contributing to sustain conservative attitudes towards embracing other business value modalities. Taking the study limitations into consideration, such as the number of providers interviewed and their geographical boundedness, it is suggested that future research continues exploring these issues both in the UK and internationally and across other types of heritage attractions to establish whether these are common across heritage attractions providers.

Despite literature suggesting that an increasingly amount of visitors may be inclined to participate in the co-creation of their experience with organisations; it appears that the majority of visitors engaging with guided tours are not interested in active interactions with heritage attractions providers, preferring more passive experiential approaches to the heritage encounter. Furthermore, findings suggest that these consumers still value their experience positively. Thus, research should focus on investigating visitor satisfaction levels (if any) among visitors taking part in co-created tours and those who experience a more passive approach (Mehmetoglu & Engen, 2011; Pine & Gilmore, 1998; 2000). Moreover, research should be carried on understanding visitor's needs, characteristics, and attitudes towards the use of co-created guided tours. Conducting such studies would help providers to better appreciate the worth of investing in customer-centric experiences and the challenges to integrate consumers into these processes (Hu & Wall, 2005; Leask, 2010; Leask & Fyall, 2006).

Disclosure statement

No potential conflict of interest was reported by the author(s).

ORCID

Iride Azara http://orcid.org/0000-0002-7159-1405

References

Alves, H., Ferreira, J. J., & Fernandes, C. I. (2016). Customer's operant resources effects on co-creation activities. *Journal of Innovation & Knowledge*, *1*(2), 69–80. https://doi.org/10.1016/j.jik.2016.03.001

Andrades, L., & Dimanche, F. (2014). Co-creation of experience value: A tourist behaviour approach. In M. Chen & J. Uysal (Eds.), *Creating experience value in tourism* (pp. 95–112). CABI.

Antón, C., Camarero, C., & Garrido, M. J. (2018). Exploring the experience value of museum visitors as a co-creation process. *Current Issues in Tourism*, *21*(12), 1406–1425. https://doi.org/10.1080/13683500.2017.1373753

Arnould, E. J., & Price, L. L. (1993). River magic: Extraordinary experience and the extended service encounter. *Journal of Consumer Research*, *20*(1), 24–45. https://doi.org/10.1086/209331

Arnould, E. J., Price, L. L., & Malshe, A. (2006). Toward a cultural resource-based theory of the customer. In R. Lusch & S. Vargo (Eds.), *The service-dominant logic of marketing: Dialog, debate and directions* (pp. 91–104). M. E. Sharpe.

Arnould, E. J., Price, L. L., & Zinkhan, G. (2002). *Consumers*. McGraw-Hill.

Azara, I., & Crouch, D. (2006). La Calvacata Sarda: Performing identities in a contemporary Sardinian festival. *Festivals, Tourism and Social Change*, 32–45. https://doi.org/10.21832/9781845410490-004

Azevedo, A. (2009, November 25-26). *Designing unique and memorable experiences: Co-creation and the "surprise" factor*. Paper presented at the III Congresso Internacional de Turismo de Leiria e Oeste—2009, Leiria, Portugal.

Bartella, G. (2014). The co-creation of animal-based tourism experience. *Tourism Recreation Research*, *39*(1), 115–125. https://doi.org/10.1080/02508281.2014.11081330

Baxter, P., & Jack, S. (2008). Qualitative case study methodology: Study design and implementation for novice researchers. *The Qualitative Report*, *13*(4), 544–559. https://nsuworks.nova.edu/tqr/vol13/iss4/2

Binkhorst, E., & Den Dekker, T. (2009). Agenda for co-creation tourism experience research. *Journal of Hospitality Marketing & Management*, *18*(2-3), 311–327. https://doi.org/10.1080/19368620802594193

Bjőrk, P. (2014). Tourist experience value: Tourist experience and life satisfaction. In N. Prebensen, J. S. Chen, & M. Uysal (Eds.), *Creating experience value in tourism* (pp. 22–34). Cabi.

Boyd, S. (2002). Cultural and heritage tourism in Canada: Opportunities, principles and challenges. *Tourism and Hospitality Research*, *3*(3), 211–233. https://doi.org/10.1177/146735840200300303

Braun, V., & Clarke, V. (2006). Using thematic analysis in psychology. *Qualitative Research in Psychology*, *3*(2), 77–101. https://doi.org/10.1191/1478088706qp063oa

Braun, V., & Clarke, V. (2012). Thematic analysis. In H. Cooper, P. M. Camic, D. L. Long, A. T. Panter, D. Rindskopf, & K. J. Sher (Eds.), *APA handbook of research methods in psychology, Vol. 2. Research designs: Quantitative, qualitative, neuropsychological, and biological* (pp. 57–71). American Psychological Association.

Bryman, A. (2012). *Social research methods*. 4th edn. Oxford University Press.

Buhalis, D., Owen, R., & Pletinckx, D. (2006). Information communication technology applications for world heritage site management. In A. Leask & A. Fyall (Eds.), *Managing world heritage sites* (pp. 125–144). Elsevier.

Campos, A. C., Mendes, J., do Valle, O. P., & Scott, N. (2016). Co-creation experiences: Attention and memorability. *Journal of Travel & Tourism Marketing*, *33*(9), 1309–1336. https://doi.org/10.1080/10548408.2015.1118424

Campos, A. C., Mendes, J., do Valle, O. P., & Scott, N. (2018). Co-creation of tourist experiences: A literature review. *Current Issues in Tourism*, *21*(4), 369–400. https://doi.org/10.1080/13683500.2015.1081158

Carù, A., & Cova, B. (2003). Revisiting consumption experience: A more humble but complete view of the concept. *Marketing Theory*, *3*(2), 267–286. https://doi.org/10.1177/14705931030032004

Carù, A., & Cova, B. (2007). *Consuming experience*. Routledge.

Centre for Economics and Business Research. (2018). *The Heritage Sector in England and its impact on the economy*. [PDF] https://historicengland.org.uk/content/docs/research/ heritage-sector-england-impact-on-economy/

Chathoth, P. K., Altinay, L., Harrington, R. J., Okumus, F., & Chan, E. S. W. (2014a). Co-production versus co-creation: A process-based continuum in the hotel service context. *International Journal of Hospitality Management*, *32*(2014), 11–20. https://doi.org/10.1016/j.ijhm.2012.03.009

Chathoth, P. K., Ungson, G. R., Harrington, R. J., Altinay, L., Okumus, F., & Chan, E. S. W. (2014b). Conceptualization of value co-creation in the tourism context. In N. Prebensen, J. S. Chen, & M. Uysal (Eds.), *Creating experience value in tourism* (pp. 33–47). Cabi.

Chathoth, P. K., Ungson, G. R., Harrington, R. J., Altinay, L., Okumus, F., & Chan, E. S. W. (2018). Conceptualization of value co-creation in the tourism context. In N. Prebensen, J. S. Chen, & M. Uysal (Eds.), *Creating experience value in tourism* (pp. 31–43). Cabi.

Chen, Z. (2018). A pilot study of the co-creation experience in traditional Cantonese teahouses in Hong Kong. *Journal of Heritage Tourism*, *13*(6), 506–527. https://doi.org/10.1080/1743873X.2018.1444045

Chen, C. F., & Chen, F. S. (2010). Experience quality, perceived value, satisfaction, and behavioural intentions for heritage tourists. *Tourism Management*, *31*(1), 29–35. https://doi.org/10.1016/j.tourman.2009.02.008

Chen, J. S., Prebensen, N. K., & Uysal, M. (2014). Dynamic drivers of tourist experiences. In N. Prebensen, J. S. Chen, & M. Uysal (Eds.), *Creating experience value in tourism* (pp. 11–21). Cabi.

Chu, T., Lin, M., & Chang, C. (2012). Mguiding (mobile guiding)—using a mobile GIS app for guiding. *Scandinavian Journal of Hospitality and Tourism*, *12*(3), 269–283. https://doi.org/10.1080/15022250.2012.724921

Chung, N., Lee, H., Kim, J. Y., & Koo, C. (2018). The role of augmented reality for experience-influenced environments: The case of cultural heritage tourism in Korea. *Journal of Travel Research*, *57*(5), 627–643. https://doi.org/10.1177/0047287517708255

Costa, C., & Buhalis, D. (2005). Introduction. In D. Buhalis & C. Costa (Eds.), *Tourism management dynamics* (pp. *Trends, management and tools* (pp. 1–6). Elsevier.

Dong, P., & Siu, N. Y. M. (2013). Servicescape elements, customer predispositions and service experience: The case of theme park visitors. *Tourism Management*, *36*, 541–551. https://doi.org/10.1016/j.tourman.2012.09.004

du Cros, H., & McKercher, B.. (2020). *Cultural tourism* (3rd ed.). Routledge.

Ferrari, S. (2013). An experiential approach to differentiating tourism offers in cultural heritage. In M. Smith & G. Richards (Eds.), *The Routledge handbook of cultural tourism* (pp. 383–388). Routledge.

Fitzsimmons, J., & Fitzsimmons, M. (2000). *New service development: Creating memorable experience*. Sage.

Fournier, S. (1998). Consumers and their brands: Developing relationship theory in consumer research. *Journal of Consumer Research*, *24*(4), 343–373. https://doi.org/10.1086/209515

Gilmore, J. H., & Pine, B. J. (2000). *Markets of one: Creating customer-unique value through mass customization*. Harvard Business School Press.

Gravari-Barbas, M., Bourdeau, L., & Robinson, M. (2016). World heritage and tourism: From opposition to coproduction. In L. Bourdeau, M. Gravari-Barbas, & M. Robinson (Eds.), *World heritage, tourism and identity: Inscription and co-production* (pp. 1–24). Routledge.

Grönroos, C. (2008). Service logic revisited: Who creates value? And who co-creates? *European Business Review*, *20*(4), 298–314. https://doi.org/10.1108/09555340810886585

Grönroos, C. (2011). Value co-creation in service logic: A critical analysis. *Marketing Theory*, *11*(3), 279–301. https://doi.org/10.1177/1470593111408177

Grönross, C., & Voima, P. (2012). Critical service logic: Making sense of value and value co-creation. *Journal of the Academy of Marketing Science*, *41*(2), 133–150. https://doi.org/10.1007/s11747-012-0308-3

Gummesson, E., & Mele, C. (2010). Marketing as value co-creationthrough network interaction and resource integration. *Journal of Business Market Management*, *4*(4), 181–198. https://doi.org/10.1007/s12087-010-0044-2

Haahti, A. (2006). *Experience design management as creation of identity economies: Reflections from periphery on entrepreneurial designs in tourism*. Environment—Sixth international conference: Entrepreneurship in United Europe—challenges and opportunities (pp. 1–15), 13–17 September 2006, Sunny Beach, Bulgaria.

Hansen, A. H., & Mossberg, L. (2017). Tour guides' performance and tourists' immersion: Facilitating consumer immersion by performing a guide plus role. *Scandinavian Journal of Hospitality and Tourism*, *17*(3), 259–278. https://doi.org/10.1080/15022250.2016.1162347

Hewison, R. (1987). The heritage industry Britain in a climate of decline.

Historic England. (2018). *Heritage counts 2018: Heritage and the economy*. [PDF] https://historicengland.org.uk/content/heritage-counts/pub/2018/heritage-and-the-economy-2018/

Hu, W., & Wall, G. (2005). Environmental management, environmental image and the competitive tourist attraction. *Journal of Sustainable Tourism*, *13*(6), 617–635. https://doi.org/10.1080/09669580508668584

Jansen-Verbeke, M., & Russo, A. P. (2008). *Innovative research on the spatial dynamics of cultural tourism*. Nova Science Publishers.

Johnston, R., & Kong, X. (2011). The customer experience: A road-map for improvement. *Managing Service Quality: An International Journal*. https://doi.org/10.1108/09604521111100225

Jonasson, M., & Scherle, N. (2012). Performing co-produced guided tours. *Scandinavian Journal of Hospitality and Tourism*, *12*(1), 55–73. https://doi.org/10.1080/15022250.2012.655078

Jung, T. H., Lee, H., Chung, N., & Tom Dieck, C. (2018). Cross-cultural differences in adopting mobile augmented reality at cultural heritage tourism sites. *International Journal of Contemporary Hospitality Management*, *30*(3), 1621–1645. https://doi.org/10.1108/IJCHM-02-2017-0084

Kim, J.-H. (2010). Determining the factors affecting the memorable nature of travel experiences. *Journal of Travel & Tourism Marketing*, *27*(8), 780–796. https://doi.org/10.1080/10548408.2010.526897

Kim, J.-H., Ritchie, J. R., & McCormick, B. (2012). Development of a scale to measure memorable tourism experiences. *Journal of Travel Research*, *51*(1), 12–25. https://doi.org/10.1177/0047287510385467

Kwortnik, R. J. (2008). Shipscape influence on the leisure cruise experience. *International Journal of Culture, Tourism, and Hospitality Research*, *2*(4), 289–311. https://doi.org/10.1108/17506180810908961

Larsen, S. (2007). Aspects of a psychology of the tourist experience. *Scandinavian Journal of Hospitality and Tourism*, *7*(1), 7–18. https://doi.org/10.1080/15022250701226014

Larsen, J., & Meged, J. W. (2013). Tourists co-producing guided tours. *Scandinavian Journal of Hospitality and Tourism*, *13*(2), 1–15. https://doi.org/10.1080/15022250.2013.796227

Leask, A. (2010). Progress in visitor attraction research: Towards more effective management. *Tourism Management*, *31*(2), 155–166. https://doi.org/10.1016/j.tourman.2009.09.004

Leask, A., & Fyall, A. (2006). *Managing world heritage sites*. Routledge.

Lee, S. J. (2017). A review of audio guides in the era of smart tourism. *Information Systems Frontiers*, *19*(4), 705–715. https://doi.org/10.1007/s10796-016-9666-6

Light, D. (2015). Heritage and tourism. In E. Waterton & S. Watson (Eds.), *The Palgrave handbook of contemporary heritage research* (pp. 144–158). Palgrave Macmillan.

Lovelock, C. H., & Young, R. F. (1979). Look to consumers to increase productivity. *Harvard Business Review*, *57*(3), 168–178.

Lusch, R. F., & Vargo, S. L. (2009). Service-dominant logic? Aguiding framework for inbound marketing. *Marketing Review St. Gallen*, *26*(6), 6–10. https://doi.org/10.1007/s11621-009-0094-6

Lusch, R. F., Vargo, S. L., & Wessels, G. (2008). Toward a conceptual foundation for service science: Contributions from service-dominant logic. *IBM Systems Journal*, *47*(1), 5–14. https://doi.org/10.1147/sj.471.0005

MacLeod, N., Hayes, D., & Slater, A. (2009). Reading the landscape: The development of a typology of literary trails that incorporate and experiential design perspective. *Journal of Hospitality Marketing and Management*, *18*(2-3), 154–172. https://doi.org/10.1080/19368620802590183

Mason, P. (2005). Visitor management in protected areas: From 'Hard' to 'Soft' approaches. *Current Issues in Tourism*, *8*(2&3), 181–194. https://doi.org/10.1080/13683500508668213

Mathisen, L. (2013). Staging natural environments: A performance perspective. *Advances in Hospitality and Leisure*, *9*, 163–183. https://doi.org/10.1108/S1745-3542(2013)0000009012

McIntyre, C. (2010). Designing museum and gallery shops as integral, co-creative retail spaces within the overall visitor experience. *Museum Management and Curatorship*, *25*(2), 181–198. https://doi.org/10.1080/09647771003737299

Mehmetoglu, M., & Engen, M. (2011). Pine and Gilmore's concept of experience economy and its dimensions: An empirical examination in tourism. *Journal of Quality Assurance in Hospitality & Tourism*, *12*(4), 237–255. https://doi.org/10.1080/1528008X.2011.541847

Melpignano, C., & Azara, I. (2018). Conserving Italian World Heritage Sites through live music events: Exploring barriers and opportunities. *Event Management*, *23*(4), 641–654. https://doi.org/10.3727/152599519X15506259855788

Milman, A. (2001). The future of the theme park and attraction industry: A management perspective. *Journal of Travel Research*, *40*(2), 139–147. https://doi.org/10.1177/004728750104000204

Minkiewicz, J., Evans, J., & Bridson, K. (2013). How do consumers co-create their experiences? *An Exploration in the Heritage Sector. Journal of Marketing Management*, *30*(1-2), 30–59. https://doi.org/10.1080/0267257X.2013.800899

Patton, M. Q. (1990). *Qualitative evaluation and research methods*. (2nd ed.). Sage.

Patton, M. Q. (2001). *Qualitative evaluation and research methods* (3rd ed.). Sage.

Payne, A. F., Storbacka, K., & Frow, P. (2008). Managing the co-creation of value. *Journal of the Academy of Marketing Science*, *36*(1), 83–96. https://doi.org/10.1007/s11747-007-0070-0

Pine, J., and Gilmore, I. I., & H, J. (2000). Satisfaction, sacrifice, surprise: Three small steps create one giant leap into the experience economy. *Strategy & Leadership*, *28*(1), 18–23. https://doi.org/10.1108/10878570010335958

Pine, B. J., & Gilmore, J. H. (1999). *The experience economy, work is theatre and every business a stage*. Harvard Business School Press.

Pine, J., & Gilmore, J. H. (1998). Welcome to the experience economy: Work is theatre and every business is a stage. *Harvard Business Review*, *76*(4), 97–105.

Potter, A. E. (2016). "She goes into character as the lady of the house": tour guides, performance, and the Southern plantation. *Journal of Heritage Tourism*, *11*(3), 250–261. https://doi.org/10.1080/1743873X.2015.1100626

Prahalad, C. K., & Ramaswamy, V. (2003). The new frontier of experience innovation. *MIT Sloan Management Review*, *44*(4), 12–18.

Prahalad, C. K., & Ramaswamy, V. (2004). *The future of competition: Co-creating unique value with customers*. Harvard Business School Press.

Prebensen, N. K., & Foss, L. (2011). Coping and co-creating in tourist experiences. *International Journal of Tourism Research*, *13*(1), 54–67. https://doi.org/10.1002/jtr.799

Prebensen, N. K., Vittersø, J., & Dahl, T. I. (2013). Value co-creation significance of tourist resources. *Annals of Tourism Research*, *42*, 240–261. https://doi.org/10.1016/j.annals.2013.01.012

Prebensen, N. K., & Xie, J. (2017). Efficacy of co-creation and mastering on perceived value and satisfaction in tourists' consumption. *Tourism Management*, *60*, 166–176. https://doi.org/10.1016/j.tourman.2016.12.001

Puczkó, L. (2013). Visitor experiences in cultural spaces. In M. Smith & G. Richards (Eds.), *The Routledge handbook of cultural tourism* (pp. 389–395). Routledge.

Ramaswamy, V., & Gouillart, F. (2010). *The power of co-creation*. Free Press.

Richards, G. (2002). Tourism attraction systems: Exploring cultural behaviour. *Annals of Tourism Research*, *29*(4), 1048–1064. https://doi.org/10.1016/S0160-7383(02)00026-9

Richards, G. (2007). *Cultural tourism: Global and local perspectives*. Psychology Press.

Richards, G. (2018). Cultural tourism: A review of recent research and trends. *Journal of Hospitality and Tourism Management*, *36*, 12–21. https://doi.org/10.1016/j.jhtm.2018.03.005

Richards, G., & Wilson, J. (2006). Developing creativity in tourist experiences: A solution to the serial reproduction of culture? *Tourism Management*, *27*(6), 1209–1223. https://doi.org/10.1016/j.tourman.2005.06.002

Rihova, I., Buhalis, D., Moital, M., & Gouthro, M. B. (2014). Conceptualising customer-to-customer value co-creation in tourism. *International Journal of Tourism Research*, *17*(4), 356–363. https://doi.org/10.1002/jtr.1993

Salazar, N. (2009). Imaged or imagined? Cultural representations and the "tourismification" of peoples and places. *Cahiers D'etudes Africaine*, *193*(19), 49–71. https://doi.org/10.4000/etudesafricaines.18628

Scherle, N., & Kung, H. (2010). *Cosmopolitans of the twenty-first Century? Conceptualising tour guides as intercultural mediators*. Paper presented at the first international research Forum on guided tours, 23–25 April 2009, Halmstad University, Sweden.

Schmitt, B. (1999). Experiential marketing. *Journal of Marketing Management, 15*(1-3), 53–67. https://doi.org/10.1362/026725799784870496

Silverman, D. (2016). *Qualitative research*. Sage.

Smith, M. K. (2016). *Issues in cultural tourism studies*. 2nd ed. Routledge.

Smith, M. K., & Richards, G.2013). *The Routledge handbook of cultural tourism*. Routledge.

Tan, S. K., Kung, S. F., & Luh, D. B. (2013). A model of 'creative experience' in creative tourism. *Annals of Tourism Research, 41*, 153–174. https://doi.org/10.1016/j.annals.2012.12.002

Timothy, D. J. (2011). *Cultural heritage and tourism: An introduction*. Channel View Publications.

United Nations World Tourism Organization. (2018). *Tourism and cultural synergies*. https://doi.org/10.18111/9789284418978.

Urry, J., & Larsen, J. (2011). *The tourist gaze 3.0*. Sage.

Vargo, S. L., & Lusch, R. F. (2006). Service-dominant logic: What it is, what it is not, what it might be. In R. F. Lusch & S. L. Vargo (Eds.), *The service-dominant logic of marketing: Dialog, debate, and directions* (pp. 43–56). ME Sharpe.

Vargo, S. L., & Lusch, R. F. (2012). The nature and understanding of value: A service-dominant logic perspective. In S. L. Vargo & R. F. Lusch (Eds.), *Special issue—Toward a better understanding of the role of value in markets and marketing (review of marketing research, Volume 9)* (pp. 1–12). Emerald Group.

Vargo, S. L., & Lusch, R. L. (2004). The four service marketing myths. Remnants of a goods-based, manufacturing model. *Journal of Service Research, 6*(4), 324–335. https://doi.org/10.1177/1094670503262946

Vargo, S. L., & Lusch, R. L. (2008). Service-dominant logic: Continuing the evolution. *Journal of the Academy of Marketing Science, 36*(1), 1–10. https://doi.org/10.1007/s11747-007-0069-6

Vargo, S. L., Maglio, P. P., & Akaka, M. A. (2008). On value and value co-creation: A service systems and service logic perspective. *European Management Journal, 26*(3), 145–152. https://doi.org/10.1016/j.emj.2008.04.003

Visit Britain. (2016). *Discover England: Summary insights on overseas visitors to England's regions*. [PDF] https://www.visitbritain.org/sites/default/files/vb-corporate/Documents-Library/documents/England-documents/discover_england_activities_reportv5.pdf

Visit Britain. (2017). Annual survey. https://www.visitbritain.org/annual-survey-visits-visitor-attractions-latest-results

Walls, A. R., Okumus, F., Wang, Y., & Kwun, D. J.-W. (2011). An epistemological view of consumer experiences. *International Journal of Hospitality Management, 30*(1), 10–21. https://doi.org/10.1016/j.ijhm.2010.03.008

Weiler, B., & Black, R. (2015). The changing face of the tour guide: One-way communicator to choreographer to co-creator of the tourist experience. *Tourism Recreation Research, 40*(3), 364–378. https://doi.org/10.1080/02508281.2015.1083742

Yeoman, I. (2012). *2050—Tomorrow's tourism*. Channel View.

Zatori, A. (2013). The impact of the experience management perspective on tour providers. In D. Koerts & P. Smith (Eds.), *3rd international research forum on guided tours* (pp. 125–137). NHTV Breda University of Applied Sciences.

Zátori, A. (2016). Exploring the value co-creation process on guided tours (the 'AIM-model') and the experience-centric management approach. *International Journal of Culture, Tourism and Hospitality Research, 10*(4), 377–395. https://doi.org/10.1108/IJCTHR-09-2015-0098

Zillinger, M., Jonasson, M., & Adolfsson, P. (2012). Guided tours and tourism. *Scandinavian Journal of Hospitality and Tourism, 12*(1), 1–7. https://doi.org/10.1080/15022250.2012.660314

Zouni, G., & Kouremenos, A. (2008). Do tourism providers know their visitors? An investigation of tourism experience at a destination. *Tourism and Hospitality Research, 8*(4), 282–297. doi:10.1057/thr.2008.30

Tourist–Tourist Social Interaction in the Co-creation and Co-destruction of Tourism Experiences among Chinese Outbound Tourists

Xing Han, Carolus L. C. Praet and Liyong Wang

ABSTRACT

The notion of tourists as co-creators of value (experiences) has been explored in the context of co-creation with service providers and residents. However, co-creation of value among tourists and the role of social interaction remain relatively underexplored. Similarly, the few extant studies on negative value creation (co-destruction) have predominantly focused on tourist–host interaction. This study examines how tourists from China visiting Japan perceive the role of social interaction with other tourists in shaping tourism experiences. Analysis of 29 in-depth interviews shows that indirect interaction with other tourists has a stronger influence on the tourism experience than direct interaction does. Perceived roles of other tourists ranged from positive (co-creation) to negative (co-destruction). The types of co-created value (Reichenberger, 2017) were mainly practical and atmospheric, whereas co-destructed value was mainly atmospheric in nature. This study also extends, confirms and partly modifies Pearce's (2005a) typology of other tourists in an East-Asian context.

Introduction

The notion of tourists as co-creators, instead of passive recipients of experiences, is now widely accepted in the field of tourism studies. Some scholars (e.g. Binkhorst & Den Dekker, 2009; Prahalad & Ramaswamy, 2004) even argue that it is the very experience of the co-creation itself that each individual consumer desires and attaches value to.

Previous research has been largely concerned with co-created experiences between tourists and service providers/organizations (Campos et al., 2016; Rihova et al., 2015), leaving the cooperative creation of experiences between tourists relatively underexplored. Nevertheless, tourism experiences[1] often take place in the presence of and/or in collaboration with other tourists. The very notion of co-creation assumes social interaction among co-creators of value as a necessary precondition for value to be co-created. The social interactions (both direct and indirect) and shared experiences with other tourists are likely to be an important aspect of the tourism experience. Indeed, some studies have addressed tourist–tourist interaction from the perspective of "co-creation" (Reichenberger, 2017; Scott et al., 2009). The co-creation of experiences between tourists involves active involvement, which results in higher levels of satisfaction and loyalty (Campos et al., 2015; Mathis et al., 2016).

While previous studies indicate that tourist–tourist interaction results in the co-creation of positive tourism experiences (Campos et al., 2015; Mathis et al., 2016), other studies suggest that when customers create value for themselves this may intentionally or inadvertently lead to a diminishment of perceived value for other customers (McColl-Kennedy & Tombs, 2011). The negative influence of other tourists on the tourism experience has been alternatively labelled "diminishment" (McColl-Kennedy & Tombs, 2011) or "co-destruction" of tourism value (Plé & Cáceres, 2010). The ever-increasing number of tourists results in frequent interactions between tourists, and therefore makes it necessary for tourism practitioners and researchers to investigate not only positive tourist–tourist interaction, but also the potential downsides of tourist–tourist interaction.

Given the growing importance of China[2] as a source market for outbound tourists in global tourism (Li et al., 2013; UNWTO, 2019), research on Chinese tourists is still relatively rare. Researchers have studied Chinese outbound tourists from a variety of perspectives, such as the cultural and philosophical differences between China and the Western world (Baggio, 2013; Fu et al., 2012; Huang et al., 2016; Kwek & Lee, 2010; Pearce et al., 2013), Chinese tourists' characteristics and behaviour patterns (Bao et al., 2018; Cheng & Foley, 2017; Chow & Murphy, 2008; Huang et al., 2015; Kim et al., 2015; Xiang, 2013; Xie & Li, 2009), impacts of tourist stereotypes and tourist-host interactions on destination image (Liu & Tung, 2017; Yang, 2015), perceptions of other tourists' annoying behaviours among outbound tourists from China and Hong Kong (Loi & Pearce, 2015), Chinese tourists' interactions with residents (Liu & Tung, 2017), Hong Kong residents' perceptions toward Chinese tourists (Siu et al., 2013), and residents' perceptions of tourism conflict with Chinese tourists in Taiwan and Thailand (Teng, 2019). Nevertheless, there is a notable lack of academic studies on tourist–tourist value co-creation and co-destruction and on the role of social interaction in tourist–tourist value co-creation and co-destruction among Chinese outbound tourists.

To address these gaps in the literature, the purpose of this study is to examine how Chinese tourists perceive the positive and negative aspects of social interaction with other, previously unacquainted tourists in the co-creation and/or co-destruction of value while travelling. We specifically focus on tourists' social interaction and co-creation/co-destruction with previously unacquainted, "stranger" tourists (Pearce, 2005a).

As we did not want to impose any preconceived structures on how Chinese tourists perceive social contact with other tourists, we used open-ended questions to conduct interviews with Chinese travellers to Japan. We used Strauss and Corbin's (1990) grounded theory method approach to code and analyse the data to let concepts emerge from the data. After analysing the data, we discovered several similarities, but also some subtle differences to Pearce's (2005a) categorization scheme of tourist perceptions of other, "stranger" tourists. In addition, we explain some of our findings by linking them to the notion of different levels of value co-creation practices, as proposed by Rihova and her associates (2013; 2015).

Literature review

Perceptions of social interaction with other tourists

Tourism involves constant encounters with other tourists and these encounters may potentially alter the experience (Crouch et al., 2001). Recent studies suggest that the social aspect is an essential and influential element of a typical tourism experience (Anderеck et al., 2006; Cutler & Carmichael, 2010). Grove and Fisk's (1997) study on tourist experiences at theme parks found that the behaviour of other tourists, either good or bad, influences tourists' overall evaluation of the tourism experience. Tourists' experiences may be influenced by so-called protocol incidents when tourists must share time and space with each other and when they must follow expected rules of conduct (Grove & Fisk, 1997). Tourists may also establish temporary friendships with each other through what Grove and Fisk (1997) have coined "sociability incidents". Huang and Hsu (2010) studied interaction between tourists on cruise vacations and confirmed the positive effect of tourist–tourist interaction on the cruise experience and on vacation satisfaction. In contrast to the impact of direct tourist–tourist interaction, several studies point out that sometimes even the mere presence of other tourists may have an impact on the tourist experience (Grove & Fisk, 1997; Praet et al., 2015; Yagi, 2001). The importance of the presence of other tourists in the tourist experience is also emphasised by Holloway et al. (2011) in what they call the "intratourist gaze", i.e. the watching of other tourists as an inevitable aspect of being a tourist.

Thus the acknowledgement of the potential impact of other "stranger" tourists who are not members of the tourist's own travelling group (i.e. the family, friends, or colleagues the tourist is travelling with), suggests that further exploration of how tourists perceive social interactions with "stranger" tourists and into the social dynamics of how tourists co-create experiences with tourists outside of their own group, is both meaningful and necessary.

Co-creation of value through tourist-tourist interaction

Several motives for social interaction between tourists have been described in the literature. Tourist interactions occur out of the desire to exchange information (Murphy, 2001), to seek for companionship, security, and belonging (Cary, 2004; Pearce, 2005b; Rihova et al., 2013), or to cope with anxieties as "temporary strangers" in unfamiliar environments (Greenblat & Gagnon, 1983). Other studies describe possible outcomes of social interaction between tourists. Social interaction of tourists may contribute to additional enjoyment (Moore et al., 2005), social development (Tung & Ritchie, 2011), engagement in the experience (Minkiewicz et al., 2013), stimulation of thoughts, feelings, and creativity (Ballantyne et al., 2011), and generate positive appraisals and memorable experiences (Campos et al., 2016; Tung & Ritchie, 2011). Huang and Hsu's (2010) study on tourist–tourist interactions on cruise vacations reveals that the tourist–tourist interaction quality is positively linked to tourist evaluation of the quality of the cruise experience and to overall satisfaction.

Recent studies acknowledge the importance of other tourists and have started to frame tourist–tourist interaction from the perspective of co-creation (e.g. Reichenberger, 2017; Rihova et al., 2013; 2015; 2018; Scott et al., 2009).

In a study of tourist–tourist co-creation and the role of social interaction among backpackers, Reichenberger (2017) found that at the level of the communitas, i.e. the temporary community created by previously unacquainted visitors, the stronger the focus on other travellers is and the longer and more personal the social interactions are, the more complex the co-created perceived value will become. Reichenberger (2017) reports four types of co-created value through tourist–tourist interaction at the communitas level, which she labels "emotional", "entertainment-related", "atmospheric", and "practical" value. Emotional value is similar to what Rihova et al. (2018) call affective value, the positive reactions expressed by others through personal social interactions. The social contact with other fellow tourists may function as a form of social support for the tourist, and may also provide a sense of belonging. Entertainment-related value appears to be somewhat similar to the notion of social value in Rihova et al.'s (2018) study, which relates to the enrichment of novel and exciting occurrences through social interactions, and sometimes may enhance self-actualization and broaden horizons due to the exposure to new cultures and perspectives of the other tourists (Reichenberger, 2017). Atmospheric value relates to the seemingly superficial and brief interactions which are viewed as mundane and insignificant, which may exert cumulative impact on the tourism experience. This type of co-created value is commonly seen when tourists take part in events such as concerts or sporting events (Reichenberger, 2017; Rihova et al., 2018). Practical value is embodied in the exchange of travel experiences and stories, which is a common topic of communication between tourists, functions as an important source of information and may impact future travel decisions (Reichenberger, 2017).

Co-destruction of value through tourist-tourist interaction

Compared to the topic of co-creation, co-destruction has received much less attention in tourism studies. The few extant studies on tourism experience co-destruction have predominantly focused on tourist–host social practices (Camilleri & Neuhofer, 2017), tourist complaining practices on social media (Dolan et al., 2019), or the integration of information and communication technologies (ICTs) into tourist experiences (Neuhofer, 2016). Camilleri and Neuhofer's (2017) study on guest–host interaction in an Airbnb context confirms that customer value can be co-created or co-destroyed through social interactions.

To the best of our knowledge, only a handful of studies have addressed the topic of tourist–tourist co-destructive social interaction. For example, Stewart and Cole (2001) find that tourists do not necessarily appreciate frequent and intensive interactions with other tourists and that the increase of interactions between tourists may even lead to the diminishment of the tourism experience. Guthrie and Anderson (2007) report that the presence of other, "stranger" tourists can have a negative impact on the destination experience, in the form of overcrowding of attractions and a resulting lack of time to appreciate the attraction, artefacts or buildings.

In contrast to studies which suggest that customer-customer interaction results in positive value co-creation, other studies from the more general field of services marketing suggest that customer behaviours to create value for themselves may intentionally or inadvertently come at the expense of the experience/perceived value of other customers and thus lead to a diminishment of value created for these other customers (McColl-

Kennedy & Tombs, 2011). Thus, the negative influences of other customers on the service experience have been alternatively labelled "diminishment" (McColl-Kennedy & Tombs, 2011) or "co-destruction" of customer value (Plé & Cáceres, 2010).

Indeed, whereas the studies by McColl-Kennedy and Tombs (2011) and Plé and Cáceres (2010) discuss the concepts of diminishment and co-destruction of value in more general servicescapes, the study by McColl-Kennedy and Tombs (2011) includes examples of restaurant and train customers whose experiences are negatively influenced by the behaviour of other customers. The examples of hospitality and transportation services are typical tourism servicescapes and therefore the constructs of value diminishment and co-destruction can be logically extrapolated to the tourism domain.

Tourist-tourist interaction involving Chinese outbound tourists

Most studies related to social interaction involving Chinese outbound tourists have focused on aspects of tourist–host interactions, in this case with residents (Liu & Tung, 2017; Siu et al., 2013; Teng, 2019). Of the two studies on tourist–tourist interaction involving tourists from China that we found, one study addresses the topic of tourist–tourist interaction and its relationship with value creation (Yang, 2015), whereas the second study discusses value co-destruction in the form of perceptions and evaluations of undesirable or unpleasant behaviours of other tourists (Loi & Pearce, 2012). In a comparative study of tourists from Hong Kong and China in Macao, Loi and Pearce (2012) report that the unpleasant behaviours of other tourists matter more to tourists from Hong Kong than they do to Chinese tourists in Macao. It was not clear however, how "other tourists" in this study were conceptualised in this study, i.e. whether they included tourists from the same country as the respondents, from other countries, or both. In contrast, Yang (2015), who investigated the impact of tourist–tourist interactions on the co-creation of destination image and the moderating effect of interaction intensity, found that favourable tourist–tourist encounters have a stronger impact on overall evaluation of interaction quality than tourist misbehaviours do. While the respondents in Yang's study were mainly from China (51%), Hong Kong (24%), Taiwan (8%), and Korea (3%), unfortunately no separate breakdown of the findings was given for the tourists from China.

The above review of the different strains of literature on social interaction, co-creation and co-destruction in tourism reveals that few studies have addressed the issue of tourist–tourist social interaction in the co-creation and co-destruction of tourism experiences. Moreover, subjects of studies on tourist–tourist interaction, co-creation and co-destruction have been primarily Western tourists in Western destinations. Finally, there is a notable lack of academic studies on outbound Chinese tourist perceptions and evaluations of the role of social interaction in tourist–tourist value co-creation and co-destruction. The current study addresses these gaps in the literature by investigating how outbound Chinese tourists perceive the role of social interaction with other tourists and how this relates to value co-creation and co-destruction. The research context is set in Japan, the second largest tourism destination in Asia for which China is the largest source market (Kennedy & Lotus, 2015; Tan, 2018).

Methodology

Given the lack of studies on social interaction, co-creation and co-destruction pertaining to outbound Chinese tourists we did not adopt any *a priori* theoretical framework, nor did we have any specific hypotheses to test. We therefore decided to follow an inductive, exploratory approach of semi-structured in-depth interviews to collect data. Respondents were informed about the purpose of the study and were asked to provide self-selected information on social interactions with other visitors that they had experienced during their current trip. We specifically invited the respondents to talk about the social interactions they had with other tourists whom they either had not previously met (Pearce, 2005a). Interview questions pertained to direct and indirect interactions with other tourists in Japan, their perception towards other tourists and willingness to engage in direct interaction with them. In this way, it was left to the respondents to select any relevant social interactions they had experienced with other visitors. The identified interactions were contextualized per event and focused on the respondents' personal feelings and evaluation of each interaction, regardless of how critical the interaction had been. Interviews were undertaken in situ to elicit and tap into vivid emotions and recent memories (Campos et al., 2015). We conducted a total of 29 interviews involving 42 respondents in June and July of 2018, of which 19 were in a tourist information centre in a large regional city in Northern Japan and ten at a popular Shinto shrine in Tokyo. Among the 42 respondents, 37 were independent tourists and 5 tourists were on a package tour.

All interviews were conducted in Chinese by the lead author and one of the co-authors. Interviews were digitally recorded with the consent of the respondents. The lead author then transcribed all the interview recordings into verbatim text omitting the paralanguage. The co-author randomly verified 20% of the transcripts to confirm their accuracy and consistency.

We conducted the coding work with a combination of manual coding and software-aided coding. The lead author first conducted manual coding, following Strauss and Corbin's (1990) approach, which includes open coding, axial coding, and selective coding. We then imported the data into the NVivo 12 software package and re-coded with the help of its coding features. The two rounds of coding generated a mature framework. The co-author compared the coding results and the corresponding original Chinese text and confirmed the validity of the coding and the resulting framework.

Findings

Respondents reported a total of 83 interaction incidents with other tourists and these ranged from positive to negative.

Direct and indirect interaction

The reported incidents include what we have labelled direct and indirect social interaction. The naming and basic conceptualization of direct versus indirect interaction follows prior studies on tourist–tourist interaction in tourism (e.g Huang & Hsu, 2009; Huang & Hsu, 2010; Kim & Lee, 2012; Yang, 2015). While previous studies have not elaborated on how "direct" and "indirect" social interaction differ, the current study conceptualizes and expands these two types of social interaction as follows. Direct social interaction occurs when tourists acknowledge each other's presence by means of verbal or non-verbal "overt" communication. In contrast, previous research on how the mere presence of other customers in a service environment may affect the customer experience and evaluation has focused on customer density or crowding (e.g. Hui & Bateson, 1991), the public behaviour of other customers (e.g. Grove & Fisk, 1997; Martin, 1996), or the mere presence of other customers in a service setting (e.g. Kim & Lee, 2012) or as a backdrop of the tourism setting (e.g. Yagi, 2001; Yagi & Pearce, 2007). Thus, indirect social interaction occurs when tourists notice and acknowledge the presence of other tourists without any direct or "overt" communication taking place. Indirect, or "inward" interaction may thus occur only inside the mind of the tourist and does not require any overt or outward forms of communication to occur. The frequency and evaluation of direct and indirect interactions with other tourists reported by respondents are illustrated in Table 1.

Respondents reported 58 incidents of direct interactions with other tourists. It is noteworthy that respondents rated the majority of direct interactions with other tourists as having little impact on their experience: 81% of the comments described direct interactions with other tourists as having a neutral impact, whereas 17% of the reported incidents described interactions as having a positive influence, and only 2% as having a negative impact on the tourism experience.

Whereas most respondents thus appeared to be somewhat indifferent to direct interactions with other tourists, they did report a considerable number (n = 25) of indirect tourist–tourist interactions. The finding of indirect interaction with other tourists as an influencing factor confirms Yagi and Pearce's (2007) suggestion that even the mere presence of other tourists could potentially have an impact on the tourism experience. It is worth noticing that most of the indirect interactions with other tourists were rated as negative (n = 12) or neutral (n = 10), while only three tourist–tourist indirect interactions were rated as positive.

Analysis of the reported cased of indirect interaction suggests that not only do tourists care about the physical and verbal behaviours of other tourists but they also are attentive to aspects such as the nationality and travel style of other tourists. Moreover, tourists may relate the behaviours of other tourists to their own. Such "relating practice" contributes to a sense of connection in shared moments or circumstances with other tourists (Rihova et al., 2018). Even the overheard conversation of other tourists, or the observed (either good or bad) manners of other tourists may constitute memorable experiences for some tourists. The following are two illustrations of indirect interactions with other tourists.

> I was taking photographs and saw some other tourists wearing kimonos (Japanese traditional robe) or Japanese style school uniforms. It was easy to remember them.

Table 1. Frequency and evaluation of tourist–tourist interactions (N = 83).

Type of interaction	Direct interaction				Indirect interaction			
Evaluation	Positive	Neutral	Negative	Subtotal	Positive	Neutral	Negative	Subtotal
n	10	47	1	58	3	10	12	25
%	17	81	2	100	12	40	48	100

(R2, Female, FIT, 20s)[3]

> The impact is usually auditory. For example, Chinese tourists, they are too noisy ... In drug stores, guests usually come in a big group and keep shouting about the products they are buying.

(R18, Male, FIT, 20s)

The impacts of the indirect interaction are also complex and subtle. The second quote confirms one of the 40 behaviours by other tourists that outbound tourists from China perceive as annoying as reported by Loi and Pearce (2015): disturbing others in public using loud voices. The respondents in this study mostly regarded big tour groups as disturbing, which is consistent with Turley and Milliman's (2000) finding that the presence of other people may have a negative impact on customers' perception of the atmosphere of a place. Hui and Bateson's (1991) study confirmed that customer density and perceived crowding exert a significant impact on the individual's choice of service. However, sometimes the presence of other tourists can be positive. One participant reported that seeing other tourists made her feel relieved as she had not encountered any other people on the way to finding a tourism attraction and was starting to wonder whether she had been going in the wrong direction. This confirms that the presence of anonymous others may have the function of providing a "sense of physical and psychological security" (White & White, 2008, p. 47). Some respondents reported that they felt relieved and reassured when hearing people speaking in Chinese when travelling in a foreign country, whereas other respondents suggested that seeing too many Chinese tourists may dilute the exotic atmosphere in Japan. Moreover, in some cases, solo travellers reported that seeing other tourists travelling with companions intensified their own feeling of loneliness.

Perceived role of other tourists

Respondents reported a range of perceived roles of other tourists. The grounded theory method for analysing our data produced a number of categories which turned out to be very similar to those found in Pearce's (2005a) typology of perceptions of other "stranger" tourists. We therefore decided to label the categories which emerged from our data by referring to and partially adapting Pearce's categories where appropriate.

Before we describe our findings pertaining to the perceived role of other tourists it is useful to revisit Pearce's (2005a) categorization of three main categories of social players that travellers interact with in the tourism experience: (1) the self; (2) other travellers; and (3) hosts. Pearce divides the "other travellers" category into strangers, family, and friends. Strangers are then subdivided into two types: (1) strangers, familiar like the travellers themselves and (2) unknown strangers. Of specific interest here is Pearce's (2005a) categorization of tourist perceptions of the roles of the "stranger" tourist, as ranging from positive to negative (see Figure 1). Moreover, Pearce (2005a) applies the concept of in-group versus out-group to distinguish between people who belong to the tourist's in-group and those that belong to the out-group. His notion of the familiar stranger refers to fellow unknown travellers whose faces tend to become familiar because of temporary physical proximity during travel (Pearce, 1980; 2005a). Pearce (2005a) positions strangers on a bipolar axis ranging from the positive pole labelled familiar stranger to the negative

	Tourists seen as	Behaviours	
Positive views	Potential close friends	Friendly contact – intimacy Learn about and from them Learn about other culture	Familiar strangers
	Travel companion	Partner for activities Socialise, someone to be with	
	Helper	Source of information Share costs Lend a hand	
	Security guard	Look after possessions Prevent unwanted contacts	
	Stimulation	Improve atmosphere of destination Excitement "Marker" of good times	
Neutral perspective	Background scenery	Just there No impact	
Negative views	Stranger	Minor discomfort Something unfamiliar	
	Disturber	Noise source Adds to crowd Invades privacy Causes conflict – contributes to culture shock	
	Competitor	Competitor for accommodation, tickets, space, access to people and setting	Total strangers

Figure 1. The multiple perspectives tourists may have of other tourists (adapted from Pearce, 2005a, p. 112).

total stranger on the opposite pole. Total strangers are people whom the tourist either has not previously met. In a tourism context it refers to people with whom the tourist generally is not inclined to socialize. The closer one gets to the total stranger pole in Pearce's original diagram, the more negative aspects of social interaction are connotated with this concept. It is interesting to note that Pearce's conceptualization of the positive, neutral, and negative perspectives of social interaction with stranger tourists from the point of view of the tourist already implicitly incorporates the notions of co-creation, co-diminishment, and co-destruction. Moreover, Pearce conceptualizes these not as binary concepts, but as the positive and negative markers on a scale with several gradations of positive and negative between the two poles.

Yagi (2001) has extended Pearce's (2005a) concept of the "familiar stranger" in a cross-national tourism context as referring to unknown travellers of the same nationality as that of the traveller, i.e. the in-group, whereas total strangers refers to people from different countries, i.e. the out-group. This conceptualization is also applicable to our findings.

The following section discusses in more detail how our respondents perceived each of the following perceived roles of other tourists which emerged from our data: temporary companion; helper/helped; familiar stranger; total stranger; disturber; and competitor.

Temporary companion

Respondents reported that they engage in casual talk with other tourists. The topics usually involved the nationalities of the tourists and their respective lives back home, stories about previous travel, and comments on the current tourism experience. Such interactions usually took place because of the intention of the tourists in question to socialize as they were accompanying each other temporarily in the destination. This appears to be similar to what Pearce (2005a) calls a "travel companion". As the respondents in this study mostly describe their socializing with other tourists as brief, superficial, mundane and insignificant, in this study the term temporary companion is used instead.

> I went on a one-day tour in Okinawa … We [other tourists and myself] had lunch and went to the aquarium together [in a group] … We talked to each other occasionally, but [all of these conversations] were very brief.
>
> (R30, Female, FIT, 30s)

Helper/Helped

Among the direct interactions with other tourists reported by respondents, most interactions were help-related, i.e. the tourist asking for help or providing help to other tourists. Types of tourist–tourist help-related interactions included showing directions, sharing travel information, taking a photograph on behalf of the other, filling out an immigration form, borrowing/lending money, and babysitting. Accordingly, and adapted from Pearce (2005a), we have labelled this role of other tourists "helper/helped". One representative quote is from a participant who expressed appreciation towards tourist–tourist help-related interaction, and regarded other tourists as a reliable source of help.

> You know that whatever problem you meet, you will always get helped if there are other tourists beside you. … People (Tourists) come here for the same purpose and would like to treat each other as in-group members.
>
> (R6, Female, FIT, 20s)

Familiar stranger

Some respondents suggested that they were delighted to see other Chinese tourists, especially when they were travelling in an unfamiliar country (such as, in this case, Japan). One of the reasons is Chinese tourists' feeling of deep-rooted bonding with their compatriots. This perspective was especially prominent among senior respondents.

> I feel that Chinese are everywhere. Whenever I am shopping or doing something else, I always have the urge to go over and say hi to the Chinese tourists I see. I just have an amiable feeling towards them.
>
> (R3, Female, group tour, 40s)

Another reason is the convenience of communication in one's native language, which made several of our respondents feel more comfortable to ask for help or express personal feelings to their compatriots. One participant reported her frustration of having trouble understanding the Japanese staff's instruction while waiting in line in Tokyo Disneyland and commented:

> It's good to see Chinese after all.
>
> (R3, Female, group tour, 40s)

Other respondents held a positive attitude towards fellow Chinese tourists, due to the absence of cultural differences. Expressing the notion that being with Chinese tourists may avoid potential friction or even conflict in contrast to being with tourists from other cultural backgrounds, one participant suggested that he would prefer to join an organized tour exclusive for Chinese rather than joining a tour with members from different countries.

> If we were all Chinese, conflict is much less likely to occur as far as I know … However, if you were with tourists from other countries, some of them may have their own opinion towards Chinese and may express their reluctance to be with you. Then it may influence your mood when travelling … The securest way is to join an organized tour that is dedicated to Chinese, and you will see no unexpected incidents. Because when you join the English tour, the members are from all parts of the world and you may probably meet some tourists who may have hostile attitudes towards Chinese.
>
> (R24, Male, FIT, 20s)

As described above, this study adopts Yagi's (2001) labelling of "familiar stranger" as referring to unknown tourists of the same nationality as that of the tourist.

Total stranger (as a neutral part of the scenery)

Some respondents viewed the tourist–tourist interaction as having little impact on the tourism experience. These respondents reported that they did not take any potential interaction with other tourists into consideration when planning their trips, nor did they believe that a random interaction with other tourists might have a substantial influence on their trip. The following quote is a good illustration:

> Actually, other tourists have not much impact, because you cannot even expect which kind of tourists you will meet. They are not part of the destination, but only temporary visitors here, just like us.
>
> (R25, Female, FIT, 20s)

In this study, this type of perceived role is labelled "total stranger" as in the original use by Pearce (2005a). As described in the literature review, Yagi (2001) has extended this familiar stranger concept in a cross-national tourism context and conceptualizes total strangers as people from different countries, i.e. the tourist's out-group. Thus the use of the term total stranger in this study includes both the original meaning of the term as used by Pearce (2005a), i.e. people that the tourist does not know (in a neutral sense), as well as Yagi's (2001) extension of the term as meaning people from other countries than one's own country. Respondents often used the term "other tourists" in a way that it was not always clear whether this term also included other Chinese. This issue occurred to us at the time of analysing the data, but at the time of the interviews we did not ask respondents to further clarify what they meant by "other tourists".

Despite the respondents' denial of any impact from other tourists on their experience, in-depth analysis of the qualitative data revealed that other tourists still had an influence on their experience, even though they themselves were not consciously aware of this. More often, the impact of other tourists is not caused by direct interaction, but merely arises due to the presence of other tourists, or because of the tourist's stereotype towards other tourists. In this case, the influence of other tourists is so subtle that the tourists themselves appear to be almost unaware of it.

Competitor

Interestingly, respondents reported more incidents of their experience being diminished by other tourists than incidents of co-creating desirable experiences with other tourists. Some respondents complained that the large number of tourists slowed down the tourism services. This perspective towards the other tourist is similar to what Pearce (2005a) has labelled competitor. It is noteworthy that the "competitor" tourists often exert impact on the tourism experience without any direct interaction taking place.

> You just need to avoid the high season. When you go to the tax refund or somewhere else, it will be full of tourists and you need to wait for an hour to get it done.
>
> (R25, Female, FIT, 20s)

Disturber

Other respondents suggested that they did not like places with too many tourists, as they believed that the touristic places are too commercialized and lack authenticity, or they simply just did not like to see a large number of tourists. Pearce (2005a) calls this type of other tourist "disturber".

> If I am going to a tourism attraction, and if there are too many tourists there, no matter which country they are from, I would be less willing to go. Having to queue is one reason, the other reason is that I would just be irritated when seeing a lot of people for no reason.
>
> (R3, Female, group tour, 40s)

Unlike previous studies probing the presence of large numbers of tourists from the perspective of perceived crowding, this study suggests that tourists may perceive other tourists as diluting the authenticity of the destination. Such perception is especially prominent towards other tourists of the same nationality.

> I come to Japan to relax but it turns out that this place has been occupied by Chinese ... If there were many tourists, but mostly from other countries than China, I would still have such feeling, but maybe not that strongly. If I were surrounded by Japanese tourists, and I was the only Chinese tourist, I would feel much better. Because I come to Japan to escape my familiar

> environment and to relax ... If there are too many tourists, even if they were from other countries than China, I would feel that this place is not pure any more ... I would think that the place is too commercialized, and that it lacks authentic beauty.
>
> (R28, Female, FIT, 20s)

Like the impact of "competitor" tourists, the above-mentioned "disturber" tourists also influence the tourism experience merely because of their presence. In other cases, respondents perceived other tourists as disturbing because of their unpleasant interaction with them (Figure 2).

> I have once taken a group tour with other tourists I hadn't met before ... Some of them made rude jokes, which made me very uncomfortable ... I think travelling with people you don't know is quite risky.
>
> (R42, Male, FIT, 40s)

Discussion

This study by and large confirms Pearce's (2005a) conceptional framework in the sense that tourists view other stranger tourists on a continuum ranging from positive to negative. Respondents held mixed views on the roles other tourists play in influencing tourism experiences, from positive roles such as temporary companion, helper/helped, neutral roles as total strangers who are part of the scenery, to negative roles such as competitor and disturber. The difference with Pearce (2005a)'s categories of stranger tourists is that our respondents did not report the positive categories of potential close friend, travel companion, security guard, stimulation, as well as the slightly negative category of "stranger" as someone unfamiliar and a source of minor discomfort, whereas they did report the other neutral and negative categories of other tourists. One other difference with Pearce's

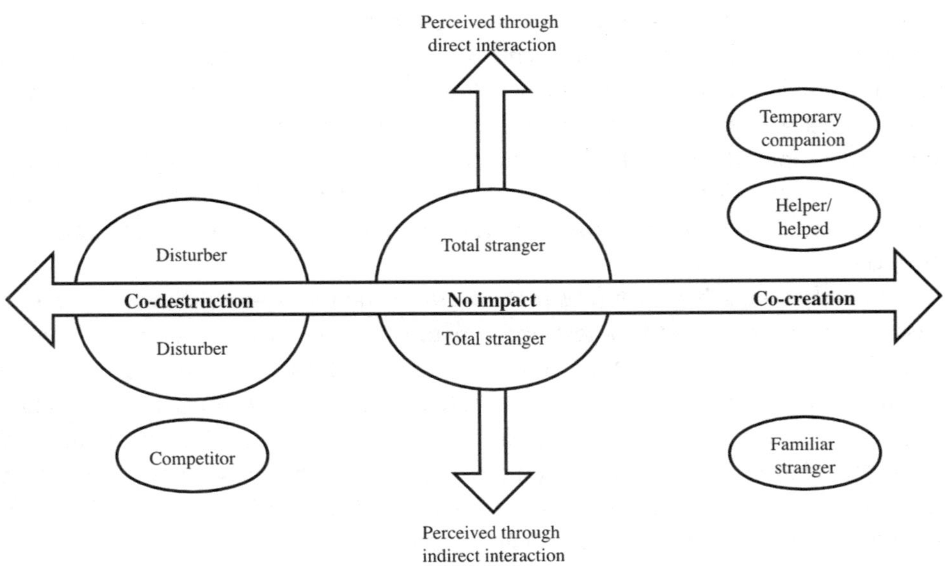

Figure 2. Perceived role of other tourists in co-creating/co-destructing the tourism experience.

scheme was that the category of other tourists as background scenery was not only perceived as neutral, but also sometimes as negative.

For some of the respondents, a positive motivation for visiting a tourism attraction was the very *absence* of other tourists. Some respondents tended to view other tourists as "competitors" for tourism resources and tourism services, or as "disturbers" of the authentic beauty of the destination. Even towards the help-related interaction with other tourists, some respondents did not hold specifically positive attitudes. The above findings suggest that Chinese outbound tourists are more inclined to create their own experiences together with the other familiar people in the semi-closed "tourist bubble" (cf. Jaakson, 2004) or "social bubble" (cf. Rihova et al., 2013; 2015) of their travel group, without the involvement of other previously unacquainted tourists. From this stance, the role of other tourists in co-creating the tourist experience is perceived as negative or at least as unnecessary. However, it is impossible for the tourist to simply ignore the presence of other tourists, as respondents reported a reasonably large number of interactions (direct or indirect) with other tourists.

An alternative way of framing the social interaction behaviours described above is to use the notion of different levels of value co-creation practices, as proposed by Rihova et al. (2013; 2015). They distinguish between the temporary communitas level, where previously unacquainted visitors create a temporary community where there is a sense of togetherness between tourists sharing an experience, the "social bubble" which refers to social practice between previously acquainted tourists, and the detached tourist level, which pertains to those who do not seek contact with unacquainted social actors. Chinese tourists appear to operate primarily at the detached tourist level of social practices, and prefer to remain in their own social bubble of people whom they already know and whom they are travelling with, whereas those that do interact socially with unacquainted tourists at the temporary communitas level (i.e. the temporary companion and the helper/helped categories described above) do so primarily at the more superficial value creation levels of what Reichenberger (2017) calls atmospheric value creation, i.e., the seemingly superficial and brief interactions which are viewed as mundane and insignificant and practical value creation of exchanging travel experiences and stories. The "familiar stranger" category of fellow Chinese whom tourists do not necessary directly interact with also has the function of co-creating atmospheric value if their presence is seen as something positive, but they may also co-destruct or co-diminish atmospheric value, if their presence is seen as negative. Perceived crowding can also be categorized as the sheer number of other tourists reducing the atmospheric value of a destination.

This study theoretically contributes to the literature on tourist–tourist social interaction and its role on co-creation and co-destruction of tourism value. The research finds that the impact of other tourists on the tourism experience of Chinese tourists is not necessarily only dependent on direct social interaction between tourists, but more often influenced through the mere presence of other tourists, i.e. indirect interaction. Moreover, this study confirms the relatively neglected phenomenon of tourists not always seeing social interaction with other tourists as something positive or, in other words, as contributing to the co-creation of tourism experiences. Instead, respondents held mixed views of the roles other tourists play in influencing tourism experiences, from positive roles (co-creator) such as temporary companion, helper/helped, neutral roles of other tourists as part of the scenery, to negative roles (co-destructor) such as competitor, and disturber. This study therefore extends, confirms and partly modifies Pearce's (2005a) typology of other tourists in the context of Chinese outbound tourism. We also identify a special segment of the other tourist, i.e. the other Chinese tourist. Chinese outbound tourists hold mixed feelings towards their compatriots when travelling. When many tourists from the same country visit the same destination, this may diminish the exotic atmosphere, whereas meeting compatriots and communicating with them in the familiar mother language reconciles the anxiety of coping with the unfamiliar environment of the foreign destination. The study also finds that practical value is the most common type of co-created value through social interaction for Chinese tourists, in addition to atmospheric value. At the same time, atmospheric value is also the one which is most commonly co-destroyed or diminished through the presence of other tourists, mainly those from China.

Our respondents reported several of the motives for social interaction between tourists that have also been identified in previous studies and that are considered to be the antecedents for value co-creation: the desire to exchange information (Murphy, 2001), to seek companionship, security, and belonging (Cary, 2004; Pearce, 2005b; Rihova et al., 2013), or to cope with anxieties as "temporary strangers" in unfamiliar environments (Greenblat & Gagnon, 1983).

In contrast, our respondents did not mention many of the possible outcomes of social interaction between tourists that have been identified by previous studies, e.g. additional enjoyment (Moore et al., 2005), social development (Tung & Ritchie, 2011), engagement in the experience (Minkiewicz et al., 2013), stimulation of thoughts, feelings, and creativity (Ballantyne et al., 2011). This, however, does not necessarily mean that these possible outcomes do not occur among Chinese travellers, only that they were perhaps not salient or important enough for our respondents to discuss or recall.

Managerial implications

This study suggests that Chinese tourists take both the possibility of co-creation as well as the avoidance of value-diminishment into consideration when choosing travel destinations and tourism activities in the destination. In contrast to findings of previous studies among Western tourists which show that social interaction with other tourists is an important motivation for travelling, our study shows that this appears to be less important for Chinese tourists. Moreover, this study reveals that some Chinese tourists have a negative perception of other tourists and see them as competitors or disturbers. It is thus necessary for tourism marketers to understand Chinese tourist perceptions towards sharing a destination with other tourists and their willingness to involve in social interactions with them.

Specifically, this study finds that Chinese tourists tend to differentiate between interactions with other Chinese tourists and those with non-Chinese tourists and that they behave differently depending on whether they are dealing with fellow Chinese or non-Chinese. It is thus necessary for tourist practitioners to have a careful understanding on how tourists from various cultural backgrounds see each other when designing sociable tourism activities, such as festivals, themed events, and group activities. Tourism marketers need to address these concerns among Chinese tourists by providing environments and activities that may facilitate tourist–tourist interactions, and by cultivating a positive expectation for tourists towards the tourism service which they will share with other tourists from various cultural backgrounds, including those from the same culture as the tourist. Tourism marketers should also consider balancing their customer mix in terms of segment compatibility and should try to avoid having too many customers from the same country at the same time and in the same place. Finally, it is also important to not treat Chinese tourists as one homogeneous market, but to understand that they are varied in their preferences and motivations for social interaction with others, with some of them not looking for contact with other Chinese, while others prefer the peace of mind of interacting with people of the same background as themselves. As this study shows, this is even true among independent travellers, who in turn may also have different attitudes towards social interaction from package group travellers. It is important for destination and hospitality managers to understand the preferences of Chinese tourists regarding the presence of tourists from China, so they can target those who prefer an environment with many other Chinese, such as those travelling on packaged group tours or alternatively target those Chinese tourists who prefer to avoid tourists from China when they visit a foreign destination. Consequently, hotels targeting organised tours or those travelling with more than one family of friends or relatives could assign rooms to tour group members on the same floor or preparing special breakfast items just for the group (Mok & Defranco, 2000), whereas hotels targeting independent travellers should perhaps ask at the time of check-in whether the guests wish to stay on a floor without other Chinese travellers, especially those travelling in large, organised tours.

Limitations and avenues for future research

Our study was based on a relatively small number of interviews and was conducted in Japan which limits the generalization of our findings. Further qualitative and quantitative studies of Chinese tourists in Japan and in other destinations are needed.

This research is only a first step towards uncovering the complexity of cooperative creation and destruction of experiences through tourist–tourist interactions among Chinese tourists. It is necessary to further examine the link between tourist evaluation of social interaction and whether the tourist was the initiator or target of the interaction (Reichenberger, 2013). It is also necessary to investigate whether communitas level social interactions are more common and lead to co-creation of emotional, entertainment-related, and self-actualization value among different types of Chinese travellers, such as backpackers. This might make it possible to confirm the existence of the "potential close friend" "stranger" tourist among Chinese travellers. In addition, further (dis)confirmation is needed of the existence of the "stimulation" and "security guard" from Pearce's (2005a) categorization scheme among Chinese travellers. More research on Chinese tourists of different age categories, generational cohorts, and travel styles is needed to further confirm whether the other types from the scheme exist among travellers from China or not.

Quantitative research is needed to verify whether it is positive tourist–tourist social interaction (Yang, 2015) or negative tourist–tourist social interaction (Loi & Pearce, 2012) that has a stronger influence on the overall evaluation of the experience among Chinese tourists. Findings of this study may be linked to other studies that suggest that Chinese tourist characteristics and behaviour are based on Chinese culture. For instance, Fu et al. (2012) claim that the influence of Confucianism causes Chinese tourists to be more cautious and distrusting toward foreign social environments but cooperative and friendly within their in-group. Indeed, this tendency to be more cooperative and friendlier towards other Chinese tourists was a theme among some of our respondents. At the same time, some respondents did not exhibit this attitude and felt that a large presence of other tourists from China had a negative impact on their experience. The evaluation of the presence of other Chinese tourists may be moderated by the number of Chinese at the destination, but also perhaps by the age/generational cohort of the tourist. Younger tourists, tourists from first-tier cities in China, more cosmopolitan so-called Transnational Chinese Post-80s tourists (Cheng & Foley, 2017) or what Arlt (2013) calls the "second wave" of New Chinese tourists consisting of "self-organised", sophisticated travellers who nevertheless still use tour operators to provide visas, flights and hotel bookings and who travel with groups of friends on relatively inflexible schedules, making them similar in their behaviour to group tourists (Xiang, 2013). Tourists who have more experience in international travel and/or experience of foreign education may be less positive towards the presence of other Chinese tourists in foreign destinations than tourists who have less experience with foreign travel or who are from second-tier and third-tier cities. Cheng and Foley (2017) argue that the generation of Chinese Post-80s tourists have their own unique value systems characterised by a clearer sense of their own rights and entitlement, the pursuit of self-identify and self-expression, with ongoing influences from Chinese socialist ideology and traditional values. While most of our respondents were travelling with family and friends and not in organised tour packages, it is not clear whether this makes them different from those travelling on organised group tours in terms of their attitudes towards interaction with fellow Chinese "stranger" tourists. One additional limitation of this study is that it was not always clear whether the term "other tourists" used by some of our respondents referred to "stranger" tourists of all nationalities (including China) or whether it only referred to tourists who were not from China. Only when respondents specifically referred to "other Chinese" this was clear. Further research is needed to explore to what extent the distinction of in-group versus out-group is based on nationality criteria, how important this distinction is in the decision to socially interact with other "stranger" tourists, and whether this differs across generations and travel styles (independent versus group tourists). Also, more research is needed on whether attitudes toward social interaction differ for "expected" interaction with "stranger" tourists at famous sightseeing spots versus "unexpected" interaction at places that are less famous or popular. Moreover, the impact of frequency and intensity of negative tourist-tourist interaction (Yang, 2015) on the overall evaluation of value co-destruction warrants further study. Finally, further research is also needed into the extent to which co-creation (positive interaction) and co-destruction (negative interaction) take place among tourists from the same and different cultural backgrounds.

Notes

1. In this paper, the terms "tourist experience" and "tourism experience" are used interchangeably.
2. In this paper, we use the terms "China" and "Chinese" to refer to mainland China.
3. We have numbered respondent quotations as R1, R2, … , where "R" means Respondent, followed by the identification number of each respondent in our database.

Acknowledgement

Xing Han acknowledges the Fuji Xerox Kobayashi Fund Research Grant Program for Foreign Doctoral Candidates in Japan, and Toshimi Otsuka Scholarship Foundation (Japan) for the financial assistance during this study.

Disclosure statement

No potential conflict of interest was reported by the author(s).

Funding

This work was supported by JSPS KAKENHI Grant: [Grant Number 18K11864].

References

Andereck, K., Bricker, K. S., Kerstetter, D., & Nickerson, N. P. (2006). Connecting experiences to quality: Understanding the meanings behind visitors' experiences. In G. Jennings, & N. P. Nickerson (Eds.), *Quality tourism experiences* (pp. 81–98). Elsevier Butterworth-Heinemann.

Arlt, W. G. (2013). The second wave of Chinese outbound tourism. *Tourism Planning & Development, 10*(2), 126–133. https://doi.org/10.1080/21568316.2013.800350

Baggio, R. (2013). Oriental and occidental approaches to complex tourism systems. *Tourism Planning & Development, 10*(2), 217–227. https://doi.org/10.1080/21568316.2013.783731

Ballantyne, R., Packer, J., & Falk, J. (2011). Visitors' learning for environmental sustainability: Testing short and long-term impacts of wildlife tourism experiences using structural equation modelling. *Tourism Management, 32*(6), 1243–1252. https://doi.org/10.1016/j.tourman.2010.11.003

Bao, J., Jin, X., & Weaver, D. (2018). Profiling the elite middle-age Chinese outbound travellers: A 3rd wave? *Current Issues in Tourism, 22*(5), 561–574. https://doi.org/10.1080/13683500.2018.1449817

Binkhorst, E., & Den Dekker, T. (2009). Agenda for co-creation tourism experience research. *Journal of Hospitality Marketing & Management, 18*(2-3), 311–327. https://doi.org/10.1080/19368620802594193

Camilleri, J., & Neuhofer, B. (2017). Value co-creation and co-destruction in the Airbnb sharing economy. *International Journal of Contemporary Hospitality Management, 29*(9), 2322–2340. https://doi.org/10.1108/IJCHM-09-2016-0492

Campos, A. C., Mendes, J., do Valle, P. O., & Scott, N. (2015). Co-creation of tourist experiences: A literature review. *Current Issues in Tourism, 21*(4), 369–400. https://doi.org/10.1080/13683500.2015.1081158

Campos, A. C., Mendes, J., do Valle, P. O., & Scott, N. (2016). Co-creation experiences: Attention and memorability. *Journal of Travel and Tourism Marketing, 33*(9), 1309–1336. https://doi.org/10.1080/10548408.2015.1118424

Cary, S. H. (2004). The tourist moment. *Annals of Tourism Research, 31*(1), 61–77. https://doi.org/10.1016/j.annals.2003.03.001

Cheng, M., & Foley, C. (2017). Understanding the distinctiveness of Chinese Post-80s tourists through an exploration of their formative experiences. *Current Issues in Tourism, 21*(11), 1312–1328. https://doi.org/10.1080/13683500.2017.1406462

Chow, I., & Murphy, P. (2008). Travel activity preferences of Chinese outbound tourists for overseas destinations. *Journal of Hospitality & Leisure Marketing, 16*(1-2), 61–80. https://doi.org/10.1080/10507050802096885

Crouch, D., Aronsson, L., & Wahlström, L. (2001). Tourist encounters. *Tourist Studies, 1*(3), 253–270. https://doi.org/10.1177/146879760100100303

Cutler, S. Q., & Carmichael, B. (2010). The dimensions of the tourist experience. In M. Morgan, P. Lugosi, & B. Ritchie (Eds.), *The tourism and Leisure experience: Consumer and Managerial perspectives* (pp. 3–26). Channel View Publications.

Dolan, R., Seo, Y., & Kemper, J. (2019). Complaining practices on social media in tourism: A value co-creation and co-destruction perspective. *Tourism Management*, *73*, 35–45. https://doi.org/10.1016/j.tourman.2019.01.017

Fu, X., Lehto, X. Y., & Cai, L. A. (2012). Culture-based interpretation of vacation consumption. *Journal of China Tourism Research*, *8*(3), 320–333. https://doi.org/10.1080/19388160.2012.704250

Greenblat, C. S., & Gagnon, J. H. (1983). Temporary strangers: Travel and tourism from a sociological perspective. *Sociological Perspectives*, *26*(1), 89–110. https://doi.org/10.2307/1389161

Grove, S. J., & Fisk, R. P. (1997). The impact of other customers on service experiences: A critical incident examination of "getting along". *Journal of Retailing*, *73*(1), 63–85. https://doi.org/10.1016/S0022-4359(97)90015-4

Guthrie, C., & Anderson, A. (2007). Tourists on tourists: The impact of other people on destination experience. In J. Tribe, & D. Airey (Eds.), *Tourism research: New directions, Challenges and Applications* (pp. 143–154). Routledge.

Holloway, D., Green, L., & Holloway, D. (2011). The intratourist gaze: Grey nomads and 'other tourists'. *Tourist Studies*, *11*(3), 235–252. https://doi.org/10.1177/1468797611432043

Huang, J., & Hsu, C. H. (2009). Interaction among fellow cruise passengers: Diverse experiences and impacts. *Journal of Travel & Tourism Marketing*, *26*(5–6), 547–567. https://doi.org/10.1080/10548400903163103

Huang, J., & Hsu, C. H. (2010). The impact of customer-to-customer interaction on cruise experience and vacation satisfaction. *Journal of Travel Research*, *49*(1), 79–92. https://doi.org/10.1177/0047287509336466

Huang, S., Keating, B. W., Kriz, A., & Heung, V. (2015). Chinese outbound tourism: An epilogue. *Journal of Travel & Tourism Marketing*, *32*(1–2), 153–159. https://doi.org/10.1080/10548408.2014.986018

Huang, S., LeBlanc, J., & Choi, H. C. (2016). How do Chinese tourists differ from Caucasian tourists? An empirical study from the perspective of tourists' self-concept. *International Journal of Tourism Sciences*, *16*(4), 222–237. https://doi.org/10.1080/15980634.2016.1257868

Hui, M. K., & Bateson, J. E. (1991). Perceived control and the effects of crowding and consumer choice on the service experience. *Journal of Consumer Research*, *18*(2), 174–184. https://doi.org/10.1086/209250

Jaakson, R. (2004). Beyond the tourist bubble? *Annals of Tourism Research*, *31*(1), 44–60. https://doi.org/10.1016/j.annals.2003.08.003

Kennedy, T., & Lotus, Y. R. (2015). Japan's China-driven tourism boom. *The Diplomat*, http://thediplomat.com/ 2015/06/japans-china-driven-tourism-boom/

Kim, N., & Lee, M. (2012). Other customers in a service encounter: Examining the effect in a restaurant setting. *Journal of Services Marketing*, *26*(1), 27–40. https://doi.org/10.1108/08876041211199706

Kim, S. S., Wan, Y. K. P., & Pan, S. (2015). Differences in tourist attitude and behavior between Mainland Chinese and Taiwanese tourists. *Journal of Travel & Tourism Marketing*, *32*(1-2), 100–119. https://doi.org/10.1080/10548408.2014.986015

Kwek, A., & Lee, Y.-S. (2010). Chinese tourists and Confucianism. *Asia Pacific Journal of Tourism Research*, *15*(2), 129–141. https://doi.org/10.1080/10941661003629946

Li, X., Meng, F., Uysal, M., & Mihalik, B. (2013). Understanding China's long-haul outbound travel market: An overlapped segmentation approach. *Journal of Business Research*, *66*(6), 786–793. https://doi.org/10.1016/j.jbusres.2011.09.019

Liu, Z., & Tung, V. W. S. (2017). The influence of stereotypes and host–tourist interactions on post-travel destination image and evaluations of residents. *Journal of China Tourism Research*, *13*(4), 321–337. https://doi.org/10.1080/19388160.2017.1399952

Loi, K. I., & Pearce, P. L. (2012). Annoying tourist behaviors: Perspectives of hosts and tourists in Macao. *Journal of China Tourism Research*, *8*(4), 395–416. https://doi.org/10.1080/19388160.2012.729411

Loi, K. I., & Pearce, P. L. (2015). Exploring perceived tensions arising from tourist behaviors in a Chinese context. *Journal of Travel & Tourism Marketing*, *32*(1-2), 65–79. https://doi.org/10.1080/10548408.2014.986013

Martin, C. L. (1996). Consumer-to-consumer relationships: Satisfaction with other consumers' public behavior. *Journal of Consumer Affairs*, *30*(1), 146–169. https://doi.org/10.1111/j.1745-6606.1996.tb00729.x

Mathis, E. F., Kim, H., Uysal, M., Sirgy, J. M., & Prebensen, N. K. (2016). The effect of co-creation experience on outcome variable. *Annals of Tourism Research*, *57*, 62–75. https://doi.org/10.1016/j.annals.2015.11.023 doi:10.1016/j.annals.2015.11.023

McColl-Kennedy, J. R., & Tombs, A. (2011). When customer value co-creation diminishes value for other customers deliberately or inadvertently. In Naples Forum on Service-Service Dominant Logic, Network & System Theory and Service Science: Integrating Three Perspectives for a New Service Agenda, 14–17 June, Capri, Italy.

Minkiewicz, J., Evans, J., & Bridson, K. (2013). How do consumers co-create their experiences? An exploration in the heritage sector. *Journal of Marketing Management*, *30*(1–2), 30–59. https://doi.org/10.1080/0267257x.2013.800899

Mok, C., & Defranco, A. L. (2000). Chinese cultural values: Their implications for travel and tourism marketing. *Journal of Travel & Tourism Marketing*, *8*(2), 99–114. https://doi.org/10.1300/j073v08n02_07

Moore, R., Moore, M. L., & Capella, M. (2005). The impact of customer-to-customer interactions in a high personal contact service setting. *Journal of Services Marketing*, *19*(7), 482–491. https://doi.org/10.1108/08876040510625981

Murphy, L. (2001). Exploring social interactions of backpackers. *Annals of Tourism Research*. *28*(1), 50–67. https://doi.org/10.1016/S0160-7383(00)00003-7

Neuhofer, B. (2016). Value co-creation and co-destruction in connected tourist experiences. In A. Inversini, & R. Schegg (Eds.), *Information and Communication Technologies in Tourism* (pp. 779–792). https://doi.org/10.1007/978-3-319-28231-2_56.

Pearce, P. L. (1980). Strangers, travellers and greyhound terminals: A study of small-scale helping behaviours. *Journal of Personality and Social Psychology*, *38*(6), 935–940. https://doi.org/10.1037/0022-3514.38.6.935

Pearce, P. L. (2005a). The role of relationships in the tourist experience. In W. F. Theobald (Ed.), *Global tourism (third Edition)* (pp. 103–122). Butterworth- Heinemann. doi:10.1016/B978-0-7506-7789-9.50012-1.

Pearce, P. L. (2005b). *Tourist Behavior: Themes and Conceptual Schemes*. Channel View Publications.

Pearce, P. L., Wu, M.-Y., & Osmond, A. (2013). Puzzles in understanding Chinese tourist behaviour: Towards a Triple-C gaze. *Tourism Recreation Research*, *38*(2), 145–157. https://doi.org/10.1080/02508281.2013.11081741

Plé, L., & Cáceres, R. C. (2010). Not always co-creation: Introducing interactional co-destruction of value in service-dominant logic. *Journal of Services Marketing*, *24*(6), 430–437. https://doi.org/10.1108/08876041011072546

Praet, C. L. C., Gotoh, H., Miyazaki, Y., Lee, J., & Wang, L. (2015). Familiar and total strangers: An exploratory study of in-group and out-group perceptions among international ski resort visitors. *Cross-Cultural Marketing in Asia, November*, 120–135.

Prahalad, C. K., & Ramaswamy, V. (2004). Co-creation experiences: The next practice in value creation. *Journal of Interactive Marketing*, *18*(3), 5–14. https://doi.org/10.1002/dir.20015

Reichenberger, I. (2013). A framework for social out-group interactions in tourism settings. In J. Fountain, & K. Moore (Eds.), *CAUTHE 2013: Tourism and global Change: On the Edge of something Big* (pp. 687–692). Lincoln University.

Reichenberger, I. (2017). C2C value co-creation through social interactions in tourism. *International Journal of Tourism Research*, *19*(6), 629–638. https://doi.org/10.1002/jtr.2135

Rihova, I., Buhalis, D., Gouthro, M. B., & Moital, M. (2018). Customer-to-customer co-creation practices in tourism: Lessons from customer-dominant logic. *Tourism Management*, *67*, 362–375. https://doi.org/10.1016/j.tourman.2018.02.010

Rihova, I., Buhalis, D., Moital, M., & Gouthro, M. B. (2013). Social layers of customer-to-customer value co-creation. *Journal of Service Management*, *24*(5), 553–566. https://doi.org/10.1108/josm-04-2013-0092

Rihova, I., Buhalis, D., Moital, M., & Gouthro, M. B. (2015). Conceptualising customer-to-customer value co-creation in tourism. *International Journal of Tourism Research*, *17*(4), 356–363. https://doi.org/10.1002/jtr.1993

Scott, N., Laws, E., & Boksberger, P. (2009). The marketing of hospitality and leisure experiences. *Journal of Hospitality and Leisure Marketing*, *18*(2-3), 99–110. https://doi.org/10.1080/19368620802590126

Siu, G., Lee, L. Y. S., & Leung, D. (2013). Residents' perceptions toward the "Chinese tourists' wave" in Hong Kong: An exploratory study. *Asia Pacific Journal of Tourism Research*, *18*(5), 446–463. https://doi.org/10.1080/10941665.2012.665062

Stewart, W. P., & Cole, D. N. (2001). Number of encounters and experience quality in Grand Canyon back-country: Consistently negative and weak relationships. *Journal of Leisure Research*, *33*(1), 106–120. https://doi.org/10.1080/00222216.2001.11949933

Strauss, A. L., & Corbin, J. M. (1990). Grounded theory research: Procedures, canons, and evaluative criteria. *Qualitative Sociology*, *13*(1), 3–21. https://doi.org/10.1007/BF00988593

Tan, H. (2018). Japan welcomed 20% more tourists in 2017 - and the number is growing. *Consumer News and Business Channel*. https://www.cnbc.com/2018/03/23/japan-welcomed-20-percent-more-tourists-in-2017– and-the-number-is-growing.html

Teng, H.-Y. (2019). Residents' perceptions of tourism conflict with Chinese tourists: Does economic dependence matter? *Asia Pacific Journal of Tourism Research*, *24*(9), 978–991. https://doi.org/10.1080/10941665.2019.1653335

Tung, V. W. S., & Ritchie, J. R. B. (2011). Exploring the essence of memorable tourism experiences. *Annals of Tourism Research*, *38*(4), 1367–1386. https://doi.org/10.1016/j.annals.2011.03.009

Turley, L. W., & Milliman, R. E. (2000). Atmospheric effects on shopping behavior: A review of the experimental evidence. *Journal of Business Research*, *49*(2), 193–211. https://doi.org/10.1016/S0148-2963(99)00010-7

UNWTO (World Tourism Organization). (2019). *International tourism Highlights (2019 Edition)*. https://www.e-unwto.org/doi/pdf/10.18111/9789284421152

White, N. R., & White, P. B. (2008). Travel as interaction: Encountering place and others. *Journal of Hospitality and Tourism Management*, *15*(1), 42–48. https://doi.org/10.1375/jhtm.15.42

Xiang, Y. (2013). The characteristics of independent Chinese outbound tourists. *Tourism Planning & Development*, *10*(2), 134–148. https://doi.org/10.1080/21568316.2013.783740

Xie, Y., & Li, M. (2009). Development of China's outbound tourism and the characteristics of its tourist flow. *Journal of China Tourism Research*, *5*(3), 226–242. https://doi.org/10.1080/19388160903149965

Yagi, C. (2001). How tourists see other tourists: Analysis of online travelogues. *The Journal of Tourism Studies*, *12*(2), 22–31.

Yagi, C., & Pearce, P. L. (2007). The influence of appearance and the number of people viewed on tourists' preferences for seeing other tourists. *Journal of Sustainable Tourism*, *15*(1), 28–43. https://doi.org/10.2167/jost528.0

Yang, F. X. (2015). Tourist co-created destination image. *Journal of Travel & Tourism Marketing*, *33*(4), 425–439. https://doi.org/10.1080/10548408.2015.1064063

Value Co-Creation and Co-Destruction: Considerations of Spa Servicescapes

Louise Buxton and Eleni Michopoulou

ABSTRACT

Spas are places that enable mind, body and spiritual harmony, and are therefore inextricably linked to the pursuit of health and wellbeing, as one of the most prominent forms of wellness tourism. Recent growth in the global spa industry is fuelled by increasing consumer interest in the pursuit of wellness. Concepts within the spa industry remain largely unexplored, thus, this conceptual paper aims to progress our understanding by considering opportunities for value co-creation and co-destruction in a spa context. In doing this, the paper unpacks the concept of the servicescape, explores the concept of authenticity and argues that understanding the consumption and production of experiences is central to understanding the creation of value in spa service settings.

Introduction

Spas are places of relaxation and rejuvenation, uniting the world of aesthetic treatments with relaxation techniques and products derived from nature's elements (Loureiro et al., 2013). Synonymous with mind, body and spiritual harmony, spas are inextricably linked to the pursuit of health and wellbeing as one of the most prominent forms of wellness tourism (Huang et al., 2019; Loureiro et al., 2013; Tabacchi, 2010).

Spas are arguably an ideal context in which to explore value co-creation and co-destruction, as they are immersive, sensory rich environments, which involve extended human interactions (Bushell & Sheldon, 2009; Lin & Mattila, 2018; Reitsamer, 2015). Thus, this paper unpacks the concept of the servicescape and considers the potential for value co-creation and co-destruction, via consumer and provider interactions and opportunities to customise sensory aspects of the spa servicescape (Bolton et al., 2014; Vargo & Lusch, 2004; Zatori et al., 2018). Furthermore, the paper explores the concept of authenticity and the notion that consumers desire unique encounters in order to generate experiential value and memorable experiences (Bharwani & Jauhari, 2013; Lin & Mattila, 2018; Oh et al., 2007; Walls et al., 2011).

Research exploring co-creation and co-destruction in the context of spa, has both practical and theoretical implications. Practically, a greater understanding of the issue could help spa operators to redraft their service offerings, and create value not just for customers, staff members and the business owners, but for all stakeholders in the value chain. Theoretically, reviewing the processes of value co-creation within a distinct and different service context (in this case spa) may provide additional insights that could perhaps also be applicable to more traditional and well-established service contexts, such as tourism and hospitality.

The growth in the global spa industry, now estimated to be worth $119 billion, is fuelled by increasing consumer interest in the pursuit of wellness (Global Wellness Institute, 2018). Despite its growth, concepts within the spa industry remain largely unexplored. The identified growth provides impetus to respond to calls from authors to better understand unexplored concepts such as value co-creation, servicescapes and memorable experiences (Kucukusta & Guillet, 2014; Lin & Mattila, 2018; Loureiro et al., 2013; Reitsamer, 2015). Responding to these calls and to advance our understanding, this paper discusses the process of value co-creation and co-destruction and considers opportunities for value co-creation and co-destruction in a spa context. Hence, this paper contributes to the body of knowledge concerning value co-creation and co-destruction, by considering through the theoretical lenses of servicescape and authenticity how these processes occur within a setting that has been largely ignored by the tourism and leisure literature.

The concept of value

Value is considered a fundamental issue to be addressed in every marketing activity (Holbrook, 1996; Sanchez-Fernandez & Iniesta-Bonillo, 2007) and despite wide interest in the concept; value is an ambiguous term (Sanchez-Fernandez & Iniesta-Bonillo, 2007). Ziethanl (1988) defined value as the consumers overall assessment of the utility of a product, derived from perceptions of what is received and what is given. However, Ziethanl's (1988) definition positions value as a uni-dimensional construct, that can be measured simply by asking consumers to rate the value they received in making their purchases, hence, this could be deemed too simple a way of explaining value (Sanchez-Fernandez & Iniesta-Bonillo, 2007). Other definitions recognise value as multi-dimensional, such as Holbrook (1996, p. 138), who identified customer value as "an interactive relativistic preference experience". For Holbrook (1996), value reflects three dichotomic dimensions: *extrinsic* versus *intrinsic, self*-versus *other-orientated* and *active* versus *reactive,* thus, providing a typology of value with eight categories: efficiency, play, excellence, aesthetics, status, ethics, esteem and spirituality. Multi-dimensional approaches to value, consider that intangible, intrinsic and emotional factors form part of the construct, which is lacking in uni-dimensional approaches (Sanchez-Fernandez & Iniesta-Bonillo, 2007). Furthermore, Holbrook's (1996) typology is useful in progressing our understanding of how value may be co-created, as value is recognised as entailing interaction between consumers and products/services, being comparative, personal and situational, being judged in a variety of ways and lying not in the purchase but in the consumption (Holbrook, 1996).

Still, there is consensus that value is determined by the beneficiary (Cannas, 2018; Skalen et al., 2015; Vargo & Lusch, 2017). As such, value is not an appreciative judgement and organisations cannot create value; nevertheless, they can make value propositions

(Nilsson & Ballantyne, 2014). Service-dominant logic holds that organisations offer value propositions, value is co-created during interactions and value is subjectively determined by the consumer (Skalen et al., 2015; Vargo & Lusch, 2008). Value propositions support consumers value creation, thus, they create opportunities for the consumer to be better off when using the product or service (Skalen et al., 2015). Value propositions, therefore, should provide opportunities for co-creation of value such as efficiency, play, excellence, aesthetics, status, ethics, esteem and spirituality (Holbrook, 1996) or memorable experiences (Hemmington, 2007; Walls et al., 2011).

Value within experiences

Pine and Gilmore (1998, 1999) highlighted the emergence of the experience economy, recognising experiences as the fourth step in the progression of economic value following on from commodities, goods and services before them. As such, consumers no longer buy service delivery and quality; they buy experiences and memories (Hemmington, 2007; Walls et al., 2011). The spa industry aims to create experiences for their consumers, hence, it could be argued that understanding the consumption and production of experiences is central to understanding the creation of value (Barnes et al., 2019; Wuttke & Cohen, 2008).

The emergence of the experience economy rendered a change in perspective of how marketing and economic activity is understood, thus, the theory of service-dominant logic evolved to shift thinking about value from an organisation perspective to a customer centric view (Vargo & Lusch, 2004, 2017). The customer centric view progresses beyond simply being customer oriented, it requires organisations to collaborate with customers, learn from them and adapt to their individual and dynamic needs (Bolton et al., 2014; Vargo & Lusch, 2004). Service-dominant logic recognises that marketing and economic activity is best understood in respect of *service for service exchange,* rather than *goods for goods* or *goods for money*; it is the activities emanating from the service which represent the source of value not the services or the goods provided (Vargo & Lusch, 2017).

Fundamental to service-dominant logic is the relationship between the consumer, the organisation and other beneficiaries (Vargo & Lusch, 2008). Consequently, value is co-created by a range of actors, not created by one actor and delivered to another (Vargo & Lusch, 2017). Moreover, the level of consumer engagement in a service, determines whether consumers are users, participants or co-creators (Bolton et al., 2014). Encounters between employees and consumers are important for experiential value creation, as employees must interact with, understand and creatively assist customers to create experiential value; however, little is known about how these encounters co-create experiential value (Barnes et al., 2019). The term value co-creation represents an evolution in marketing thought, as it positions consumers as active players in their experience, rather than passive audiences, thus, value is co-created for consumers, through their interactions and customisation of their experiences (Bolton et al., 2014; Vargo & Lusch, 2004; Zatori et al., 2018).

However, there can be interactions whereby the exchange of resources between the collaborating parties may also result in negative experiences; often referred to as value co-destruction (Ple & Caceres, 2010). Corsaro (2020) explains that value co-destruction focusses on collective collaborations between consumers and providers, and that those actors can suffer different levels of co-destruction resulting from the interactional processes. Further explaining the concept of value co-destruction, Javri et al. (2020) suggest that consumers' previous experiences develop a specific cognitive script which identifies how providers and consumers are expected to behave. Thus, this shapes what they expect from consumption experiences. If either the provider or the consumer deviates from the script, value co-destruction, rather than co-creation may emerge (Javri et al., 2020). For example, if a consultation, prior to the guest having a spa treatment is not conducted thoroughly, when it has been on the guest's previous visit, value may be co-destructed.

Value of emotions and memorability of experiences

Connected to the evolution of economic value in the experience economy, the way consumers interact with offerings has changed. Whilst commodities, goods and services are external to the consumer, experiences are internal, created in the mind of the consumer, as they are engaged on an emotional, intellectual and often spiritual level (Bastiaansen et al., 2018; Oh et al., 2007; Pine & Gilmore, 1998, 1999; Pullman & Gross, 2004; Smith & Colgate, 2007). Using Holbrook's (1996) typology of value, it can be considered how intrinsic, self-oriented, active value categorised as "play" leading to "fun", could be co-created during a white-water rafting trip. To realise the value proposition, the organisation must co-create value with its consumers by direct interaction (Vargo & Lusch, 2008), which in this example could involve instructing the consumer how to use a paddle whilst white-water rafting (Skalen et al., 2015). It could even be a demonstration by spa staff of how to use a particular skincare product.

The focus on emotions in customer-employee encounters is argued to be central to experience value creation within tourism and hospitality (Barnes et al., 2019). Emotions are the core building blocks of experiences, thus, experiences are only memorable when they evoke emotions (Bastiaansen et al., 2018). In the experience economy, consumers want more than the delivery of services, they seek unique consumption encounters to accompany the services, which create memorable experiences (Walls et al., 2011). Hence, it is no longer enough to offer a functional level of service; consumer demand for unique and memorable experiences requires organisations to develop value-added experiential provision which connects with individuals on a personal and emotional level (Bharwani & Jauhari, 2013; Lin & Mattila, 2018; Oh et al., 2007; Walls et al., 2011).

Ample research has dealt with the concepts of satisfaction and emotions (Barsky & Nash, 2002; Bloemer & de Ruyter, 1999; McIntosh & Siggs, 2005). For instance, McIntosh and Siggs (2005) highlight that enjoyment for boutique hotel guests resonates within sensory and experiential aspects of their stay, which underlines the importance of emotions and personalised attention. Their findings also showed that travellers appear not to be concerned with just being "there" in the tourist setting, but are concerned with participating, learning and experiencing the "there" that they visit and concluded that more research emphasis needs to be directed at all integral aspects of the tourist experience.

Walls et al. (2011) identified that experiential research in tourism and hospitality settings had taken three directions: creating taxonomies of experiences, examining the cause of, or explaining experiences and comparing the relationship between experiences and other constructs. For example, Thorne (1963) explored the concept of peak experiences and provided a taxonomy of six main categories: sensual, emotional, cognitive, conative, self-actualisation and climax. Pine and Gilmore (1999) suggested there are four realms of experience: education, entertainment, escapism and aesthetics. This framework was later revisited by Oh, Fiore and Jeoung (2007), who confirmed the dimensional structure of the four realms of experience but suggested that further research could provide conceptual clarification regarding the relationship of experience economy concepts with general consumption evaluations such as memorable experiences.

Hence, creating experiential value is relevant to how organisations design appropriate experiences to include feelings and emotions (Smith & Colgate, 2007). This value-added experiential provision can be offered in the form of entertainment, education, an escapist or aesthetic experience (Oh et al., 2007; Pine & Gilmore, 1998). Examples of added value in hospitality and tourism include shopping centres offering concerts, a hotel cooking demonstration, themed guestrooms or restaurants with singing waiters, the added value for the consumer being the memorable experience (Mehmetoglu & Engen, 2011; Oh et al., 2007). It can be argued that added value in a spa setting takes the form of learning how to use skincare products through watching a demonstration, other educational programmes geared towards improving health or wellbeing, or in the form of themed spa treatments (Buxton, 2018; Lin & Mattila, 2018). Many spas also have beautiful surroundings, and value could be added through utilising the location to allow guests to take in the views and enjoy being in the destination, passively appreciating how the setting appeals to the senses.

Value of authentic experiences

At the broadest level, "authenticity is the concept aimed at capturing dimensions of truth or verification" (Newman & Smith, 2016, p. 610). Different uses of the term "authenticity" exist across many academic disciplines including aesthetics, philosophy, sociology, psychology, marketing and tourism and academics, within each discipline, have sought to construct typologies to define their focal phenomenon (Newman & Smith, 2016). Bruner (1994) and Reisinger and Steiner (2006), both noted challenges with the term authenticity. However, the term is still in use in tourism literature today, but it has evolved through reconceptualization and clarification (Checa-Gismero, 2018; Steiner & Reisinger, 2006; Tiberghien, 2019; Wang, 1999; Yang, 2019). However, a growing nomenclature exists within tourism literature, with some terms overlapping with each other. To provide clarification, attempts have been made to explain authenticity in tourism based on many factors, including the type of tourist experience (Cohen, 1979), the nature and existence of what is being judged authentic (Laing et al., 2014; Wang, 1999), the interplay between different dimensions (Belhassen et al., 2008) or ideological perspectives of authenticity (Reisinger & Steiner, 2006; Wang, 1999).

Authenticity is a recent addition to the study and language of spa (Laing et al., 2014; Lin & Mattila, 2018; Michopoulou, 2017; Poluzzi & Esposito, 2020); however, it has existed as a concept, in the area of tourism for several decades. Almost half a century ago, MacCannell (1973) introduced the concept of authenticity to sociological studies of tourists' motivations and experiences. Subsequently, authenticity became a key concept in an emergent sociological paradigm for the study of tourism (Belhassen et al., 2008; Cohen, 2007;

Wang, 1999). The term authenticity was originally used in reference to the genuineness of objects in museums, whilst it has also been used to refer to the human condition (Steiner & Reisinger, 2006). Two decades on from MacCannell's seminal work, Wang (1999) proposed that the issues of authenticity in tourism could be divided into two separate issues: the authenticity of toured objects and of experiences. These two issues can both be applicable to spa, in relation to the authenticity of the physical environment, its location, spa facilities and décor and the authenticity of the spa experience (Poluzzi & Esposito, 2020).

Highlighting further the complex nature of authenticity, Wang (1999) described three different types of authenticity: objective, constructive (or symbolic) and existential, the latter was divided into two dimensions: *intra-personal* and *inter-personal* authenticity. At the turn of the century, there was a shift in the discourse of authenticity, way from the much-contested objective and concrete view of authenticity towards the more subjective and abstract view (Belhassen et al., 2008; Cohen, 2007). Cohen (2007) recognised that this shift had been in part initiated by the theoretical difficulties of the discourse of objective authenticity and by the emergence in tourism studies of postmodern ideas, such as those of Baudrillard (1983) regarding the disappearance of "originals". The concept of existential authenticity has been the most enduring in the study of experience consumption, due to its applicability to the multi-dimensionality of experiences (Belhassen et al., 2008; Cohen, 2007; Wang, 1999). A recent illustration of this shift is provided by Le et al. (2019) who note that authenticity has been used to as a feature to lessen the homogeneity of mass produced, commoditised goods, services and experiences.

With the existent focus on the experience economy and the recognition for potential transformation through experiences (Pine & Gilmore, 1998, 2011; Pizam, 2010; Sipe & Testa, 2018), understanding authenticity in consumption experiences such as dining out (Le et al., 2019), board and lodgings (Mody et al., 2019) and even spa (Lin & Mattila, 2018; Poluzzi & Esposito, 2020) has become increasingly important. Existing conceptualisations of authenticity have largely been developed in a tourism context, therefore, there are gaps in our understanding of how authenticity applies in other experience settings. There is, in particular, limited understanding of the concept of authenticity as it applies to spa (Le et al., 2019; Lin & Mattila, 2018).

Objective (intrinsic) authenticity

Objective authenticity (also referred to as intrinsic authenticity) applies the museum-linked usage of authenticity to describe the toured objects as perceived by tourists and it follows that the authentic experience is a result of the recognition by tourists that the toured objects are authentic (Wang, 1999). Representing the modernist/realist/objectivist ideological view of authenticity (Reisinger & Steiner, 2006), the criterion to be authentic or inauthentic was objective, in that it was whether, the tourist object was made or the experience was enacted by local people according to custom and traditions (Sharpley, 2018). As such, authenticity denotes a sense of the genuine, real or uniqueness (Wang, 1999). Michopoulou (2017) suggests that intrinsic authenticity is potentially associated with spa tourism experiences, in particular spa treatments, rituals and therapies that are linked to the country of origin and culture. The sense of intrinsic authenticity in a spa might arise from tangible factors such as heritage features or décor or intangible factors such as traditions and stories linked to the spa and /or its treatments (Laing et al., 2014). For example, Rosewood Hotels and Resorts describe their spas' designs and concepts as being inspired by purity, authenticity, and focused on a sense of place (Kitchen, 2019). Furthermore, Rosewood Hotels and Resorts research local traditions and employ local healers and practitioners to develop spa treatments and rituals, which they offer as "Lost Remedies" in their spa menus (Kitchen, 2019). Similarly, two Chable resorts in Mexico use indigenous resources in their treatments which are linked to Mayan traditions (Megson, 2019).

Constructive authenticity

Wang (1999) asserts that constructive authenticity is the result of social construction, not an objective measurable quality of what is being visited, thus, there are various versions of authenticities regarding the same objects. Constructive authenticity is largely affiliated with academic schools of constructivism (Belhassen et al., 2008). From a constructivist perspective, the ontological assumption is that there is no real world that is independent of human mental activity, thus, pluralistic and relativist epistemologies and methodologies hold (Wang, 1999).

Constructivists assert that there is no absolute and static original on which the authenticity of originals relies; origins and traditions are invented and constructed. This construction involves power and is a social process. Knowledge of authenticity is created, not discovered by the mind; it is pluralistic and relative to individuals. Authenticity is a label attached to the visited people or cultures in terms of stereotyped images and is a projection of the tourists own beliefs, expectations and consciousness of those (Wang, 1999). Those stereotyped images, expectations and consciousness are all are subject to influence from mass media, movies and word of mouth from friends and family, thus, the constructivist approach to authenticity identifies that individuals perceptions coagulate into a socially constructed recognition of the authenticity of the phenomenon (Le et al., 2019). Tourists are indeed in search of the authentic, however, what they seek is not objective; it is symbolic authenticity, which is the result of social construction.

For example, spa treatments and rituals have become globally transportable (Laing et al., 2014; Michopoulou, 2017), with experiences such as Thai massage, Swedish massage and hammam treatments appearing ubiquitously on spa menus. Nonetheless, Laing et al. (2014) suggest that many consumers believe an authentic spa experience can only be gained in the country of its origin. Lin and Mattila (2018) similarly, note that many spas offer a hammam treatment and have the required tiled steam room in which to provide it, however, they cannot deliver the true essence of a hammam as delivered in Turkey, Morocco or Tunisia. It is acknowledged that consumers perceptions of authenticity will be influenced by their own culture and experiences (Lin & Mattila, 2018; Michopoulou, 2017) and the interpretation of authenticity can be multifaceted and personal (Reisinger & Steiner, 2006), so simply offering a tiled steam room may be enough for some consumers to deem the experience authentic. However, those who have experienced a hammam in Turkey, Morocco or Tunisia may not be satisfied with this, and will only view the experience authentic, if they are experiencing it in Turkey, Morocco or Tunisia (Laing et al., 2014; Lin & Mattila, 2018).

Existential authenticity

The differentiation of the issue of the "authenticity of experiences", from the issue of the "authenticity of objects", was central to the introduction of the concept of existential authenticity (Wang, 1999). Distinct from both objective and constructive authenticity, which focus on whether and how toured objects are authentic, existential authenticity involves personal feelings triggered by the liminal processes of tourist activities (Wang, 1999). The surroundings and environment of a travel experience can serve as a catalyst for existential authenticity (Brown, 2013). Wang (1999) argued that objective and constructive authenticity, as object related concepts, have limited application to a number of tourist activities, whereby, existential authenticity is fitting to explain more tourist experiences. Hence, if existential authenticity is important to examine within the realm of experiences, it is pertinent to consider it also within spa experiences.

Intra-personal authenticity

Wang (1999) further divided the concept of existential intra-personal authenticity, developing a framework, which differentiates between two aspects, "self-making" and "bodily sources of authentic self" (Laing et al., 2014). The first aspect of the intra-personal dimension of authenticity, Wang (1999) described was self-making or self-identity, through tourism experiences. For many people, the constraints and routine of everyday life make it difficult to achieve self-realisation; thus, self-making is an implicit motivation for engaging in tourist activities, in particular those which provide challenge or adventure (Wang, 1999). For the spa tourist, a spa experience may provide them with the time to contemplate who they really are (Laing et al., 2014). Alternatively, it may provide an escape from everyday life by allowing the guest to feel pampered or celebratory (Laing et al., 2014). Loureiro, Almeida, and Rita (2013) in their study of the effect of atmospheric cues and involvement on pleasure and relaxation on spa, assert that the extent to which consumers express their self-concept through involvement in spa experiences cannot be overlooked. Whilst there are many motivations to engage in spa experiences, an overarching theme exists in the desire to transform the self, for example, through engaging in wellness activities or spa therapies, whereby the consumer may feel they are a more authentic version of themselves (Laing et al., 2014; Wang, 1999). Akin to this, Rosewood Hotel and Resorts have recently introduced a spa retreat in Mexico called "Marry One Self" which they purport is a "true journey of self-acceptance and self-love" (Kitchen, 2019, p. 48). The experience includes a consultation with the resort's resident shaman followed by a tamazcal session (indigenous sweat lodge), massage and skincare (Kitchen, 2019).

The second aspect of the existential intra-personal dimension of authenticity involves bodily feelings such as relaxation, rehabilitation, recreation, excitement, sensual pleasure and play, all touristic contents (Cohen, 1979; Wang, 1999). Furthermore, Wang (1999) asserts that "bodily concern" consists of two aspects: *sensual*, bodily feelings and *symbolic*, the culture and sign system of the body. With the former, the body is not just a corporate substance but also the "feeling subject" (Seamon, 1979 cited in Wang, 1999, p. 362) with the latter, the body becomes a show of personal identity of health, beauty, fitness and youth (Wang, 1999). Laing et al. (2014) note that the bodily form of authenticity is particularly interesting in the spa context, given the appeal of spa treatments to the senses. However, the symbolic aspect is also applicable; for example, spa experiences may provide symbolic evidence, to demonstrate that the guest is engaging in an aspirational lifestyle, whatever the reality of their social status and financial circumstances may be (Laing et al., 2014).

Inter-personal authenticity

The inter-personal dimension of existential authenticity, explains the tourists' quest for authenticity through connections with others (Wang, 1999). This applies to spa experiences, as the consumer may gain a sense of authenticity via their interactions with other people; for example, travel companions, other spa guests and in particular, interactions with spa staff, through learning about their culture and heritage (Laing et al., 2014). Considering consumers' interactions with spa professionals, a spa treatment requires one-to-one interaction between the service provider and the guest; thus, the spa industry relies extensively on its employees. The nature of the spa service requires employees to demonstrate meticulous grooming, gentle gestures and communication and the skilled delivery of treatments (Lin & Mattila, 2018). If the spa professional does not appear sufficiently engaged physically and emotionally during the experience, a lack of inter-personal authenticity may be perceived (Laing et al., 2014). The service providers' and guests' cultural backgrounds may also influence the perception of authenticity (Öznalbant & Alvarez, 2020). For instance, if a Hawaiian lomi-lomi massage is delivered by a spa professional who is Hawaiian, the perception of authenticity could be increased, particularly for a guest who is not Hawaiian (Lin & Mattila, 2018).

Experiences within servicescapes

The servicescape has been widely conceptualised (Bitner, 1992; Nilsson & Ballantyne, 2014; Rosenbaum & Massiah, 2011; Tombs & McColl-Kennedy, 2003). Servicescape is the term coined to describe the environment in which a service process takes place (Bitner, 1992). More than just the physical surroundings, servicescapes include anything observable that effects consumers' perceptions and behaviours (Kim et al., 2016). Servicescapes act as the packaging of the service (Mari & Poggesi, 2011). Many stimuli within a servicescape can influence the consumer's emotional state, and these can be both tangible and intangible factors (Yang & Namkung, 2009). Holbrook (1996) explains one dichotomic dimension of value as *active*, when it involves some physical or mental manipulation of the tangible or intangible, thus, it could be argued that allowing consumers to customise an aspect of the servicescape, for example choosing the colour of lighting in a room could be a value proposition.

To explore opportunities for value co-creation with a spa context, it is important to first unpack the servicescape concept and then consider the aspects within the spa servicescape which could provide value propositions. The servicescape is the context for service, containing physical, sensory and social dimensions essential to the creation of service experiences (Bitner, 1992; Nilsson & Ballantyne, 2014). Servicescape factors have been identified as: ambient conditions, functionality, and signs, symbols and artefacts, service staff (Bitner, 1992) and other consumers (Tombs & McColl-Kennedy, 2003). Ambient conditions and sensory factors such as sound, sight, smell, taste and touch have all been shown to effect consumer behaviour (Abhishek et al., 2013; Ballantine et al., 2010; Decre & Pras, 2013; Krishna et al., 2010). Effects include increasing intentions to buy (Decre & Pras, 2013), attracting consumer attention (Ballantine et al., 2010) and increasing memory (Krishna et al., 2010). Features such as, decorations, signs, symbols and artefacts, are designed into the service environment and provide cues about the service provision to the consumer (Kim et al., 2016; Nilsson & Ballantyne, 2014). Additionally, factors such as layout and functionality facilitate the ease of the provision of service (Nilsson & Ballantyne, 2014).

Evidence of the importance of ambient and sensory conditions in a servicescape is available across different service industries (Ballantine et al., 2010; Kim et al., 2016; Loureiro et al., 2013). For example, Ballantine et al. (2010) investigating the role of atmospherics in the creation of a hedonic retail experience, found that colour, lighting, music and noise all have moderate positive effects on attracting consumer attention. Abhishek et al. (2013) assessing the role of touch in shopping, found that consumers in a retail setting judge product quality through haptic touch. Similar findings have been recorded within hospitality and tourism settings, as Kim et al. (2016) found that in leisure centres, design aspects such as decoration, accessibility and signage are important features of the service encounter. In hospitality, design, space, lighting, colour and music are also important stimuli which may influence consumer behaviour, i.e. purchase intentions or how long a consumer spends in the restaurant setting (Milliman, 1986; Yang & Namkung, 2009). Expectedly, the importance of Consumers considers the social relationships built with focal employees as a relational benefit (Rosenbaum & Massiah, 2011). Social facilitation theory draws attention to the effects on the consumption experience, of the presence of other consumers within the servicescape (Clendenen et al., 1994; Lovelock, 1996; Nakata & Kawai, 2017; Turley & Milliman, 2000) and consumers are influenced by social density (Rosenbaum & Massiah, 2011). Tombs and McColl-Kennedy (2003) explain that consumers' approach/avoidance behaviour towards crowds, was driven by their desire for private or group consumption. In hospitality, for example, the performance of service staff is essential to consumer perceptions of the service offering (Yang & Namkung, 2009), and in restaurants, staff behaviour and image can positively influence customer loyalty intensions (Harris & Ezeh, 2008). Perhaps even more crucial are the findings by Hanks and Line (2018), who suggest that the mere presence of others in the hospitality servicescape can influence the perception of consumers. In a spa context, the guest experience could be adversely impacted if a spa pool or sauna were occupied by other guests; when a guest wanted to experience them in seclusion.

Experiential value co-creation and co-destruction within spa servicescapes

Empirical investigation of the servicescape has taken place in many service settings such as leisure centres (Kim et al., 2016), restaurants (Hanks & Line, 2018; Harris & Ezeh, 2008) and retail (Abhishek et al., 2013; Ballantine et al., 2010; Decre & Pras, 2013). Despite extensive research of the impact of the servicescape on consumer behaviour (Decre & Pras, 2013; Krishna et al., 2010; Nakata & Kawai, 2017; Nilsson & Ballantyne, 2014; Rosenbaum & Massiah, 2011), investigation of the servicescape concept in spa is limited and there is a lack of research which addresses the issue of the servicescape's impact on value co-creation. Thus, exploring the concept of the servicescape and process of co-creation in a spa context could shed some light into this largely unexplored area.

Many factors within the spa servicescape arguably provide opportunities for value co-creation, for example, spa consumers spend an extensive amount of time with spa service staff during their service encounter, therefore, it could be contended that the interactions they have and relationships they build with focal employees are fundamental to the co-creation of experience value (Barnes et al., 2019; Rosenbaum & Massiah, 2011). Furthermore, customised choices of massage oil, music, room scent and lighting colours could position consumers as active players in their experience, co-creating value through customisation of the experience (Bolton et al., 2014; Vargo & Lusch, 2004; Zatori et al., 2018).

However, value co-destruction arguably could occur if there is an absence of information for a consumer to understand the spa facilities [sauna, steam room, spa pool] or lack of signage for a consumer to navigate through the environment, for example, to find the changing rooms or relaxation area. If relationships with service staff do not build sufficient levels of trust, value may be co-destructed (Rosenbaum & Massiah, 2011). If the spa consumer has previous experiences of consuming spa, thus, has a cognitive script for what they expect, value may be co-destructed, if that script is not followed by the provider (Javri et al., 2020). Likewise, spa experiences are rarely consumed in isolation from other consumers; if the spa setting is more crowded than consumers expected (Tombs & McColl-Kennedy, 2003) or if other consumers misbehave, value may be co-destructed (Corsaro, 2020).

Despite the popularity of the experience economy, there is a scant research specifically in a spa context. One notable exception is Lin and Mattila (2018, p. 42) who created a conceptual model which proposes that, in a spa setting, the components of "natural environments, the servicescape, service encounters, spa treatments: culture and rituals, and spa educational programmes geared towards health and wellbeing, combine to provide psychological, emotional and social benefits" could co-create experiential value. However, this model has not been subject to empirical testing. Empirical research in a spa setting has focused on the effects of atmospheric cues (Loureiro et al., 2013), consumer preferences (Kucukusta & Guillet, 2014), sensorimotor perceptions on consumers cognitive and behavioural responses in servicescapes (Reitsamer, 2015). Furthermore, all studies have identified the need for further research (Kucukusta & Guillet, 2014; Lin & Mattila, 2018; Loureiro et al., 2013; Reitsamer, 2015). Exploration of experience economy constructs in a spa setting could respond to these calls and those from tourism and hospitality (McIntosh & Siggs, 2005; Oh et al., 2007) to provide a greater understanding of this growing area of experiential consumption.

This research has several managerial implications that are divided into three main areas: redrafting of service offerings, training, and value chain co-creation. In respect of redrafting service offerings, this could include assisting managers to develop best practice in: *redesigning servicescapes*, by providing spa operators with a better understanding of the signage and information guests require to know how to use the spa facilities or navigate through the different spaces (Kim et al., 2016); *scheduling of guest bookings*, via discovering the maximum number of guests to allow into different spaces at one time so that they are not too crowded that value is co-destructed (Rosenbaum & Massiah, 2011); *offering opportunities for customisation* through identifying which customised choices contribute to greater value co-creation (Bolton et al., 2014), so that efforts can be focused effectively. The research could have an impact on the advancement of appropriate training for service personnel, to creatively assist them to develop the skills required to build effective relationships with their guests, which lead to value co-creation (Barnes et al., 2019; Rosenbaum & Massiah, 2011). Moreover, managers should not think narrowly in respect of value co-creation, instead of focussing purely on consumer and staff perspectives, they should take a broader, view paying attention to the entire value chain and on bringing value to all relevant stakeholders.

Conclusion

In summary, it is evident that there is growing demand for spa experiences (Global Wellness Institute, 2018) and research in a spa context could help to better understand the process of value co-creation and co-destruction for spa consumers (Nilsson & Ballantyne, 2014). There is consensus that situational, environmental and human factors contribute to the customer experience (Bitner, 1992; Corsaro, 2020; Lin & Mattila, 2018; Rosenbaum & Massiah, 2011; Tombs & McColl-Kennedy, 2003; Walls et al., 2011). Despite the important conceptual works on value co-creation, the servicescape and memorable experiences, empirical examination of the subjects in a spa context, is particular limited, with several authors calling for further research (Buxton 2018; Corsaro, 2020; Kim et al., 2016; Lin & Mattila, 2018; Mehmetoglu & Engen, 2011; Michopoulou, 2017; Oh et al., 2007; Walls et al., 2011).

More than a decade on from its inception as a lens through which to view service, Vargo and Lusch (2017) suggest that to move service-dominant logic forward, the development of more mid-range theory and evidenced based research is required. Spas provide immersive, sensory rich environments, and spa services involve extended human interactions (Bushell & Sheldon, 2009; Lin & Mattila, 2018; Reitsamer, 2015). Thus, it can be argued spas provide ripe opportunities to explore value co-creation and co-destruction. Mid-range theory could take the form of developing a framework of factors in a spa context that may impact on value co-creation and co-destruction. The development of such a framework may provide a structure to enable empirical investigation of value co-creation and co-destruction in a spa context. This would have practical implications as it can be shared with spa operators to inform the design of spa offerings, development of training and the enhancement of value propositions. Furthermore, the development of a framework could help to better understand value co-creation and co-destruction in the context of spa and provide opportunities to test methods of investigation that have been applied in other contexts, such as tourism and hospitality, in a new context.

This research has limitations which could be addressed with further research. Although the conceptualisation provides a good understanding of factors which could contribute to co-creation and co-destruction of value in a spa context, it has not been tested and therefore lacks empirical data, which would provide greater insight. The research addresses the issue of value co-creation and co-destruction through the lens of the servicescape and authenticity; in further research, this view could be expanded to include other perspectives. Attention is given specifically to value co-creation and co-destruction for consumer and spa operators, further studies could delve deeper to understand this for a broader range of actors across the whole value, supply and distribution chain. The research was developed with hotel, resort and destination spas in mind, research exploring how value is co-created on co-destructed in different spa servicescapes such as mineral spring spas or medical spas would provide insights into co-creation and co-destruction in different settings.

Disclosure statement

No potential conflict of interest was reported by the author(s).

ORCID

Eleni Michopoulou http://orcid.org/0000-0002-1857-4462

References

Abhishek, S., Sinha, P. K., & Vohra, N. (2013). Role of haptic touch in shopping. *Decision*, *40*(3), 153–163. https://doi.org/10.1007/s40622-013-0017-x

Ballantine, P. W., Jack, R., & Parsons, A. G. (2010). Atmospheric cues and their effect on the hedonic retail experience. *The International Journal of Retail and Distribution Management*, *38*(8), 641–653. https://doi.org/10.1108/09590551011057453

Barnes, S. J., Mattsson, J., Flemming, S., & Jensen, J. F. (2019). The mediating effect of experiential value on tourist outcomes from encounter-based experiences. *Journal of Travel Research*, *59*(2), 367–380. https://doi.org/10.1177/0047287519837386

Barsky, J., & Nash, L. (2002). Evoking emotion—affective keys to hotel loyalty. *Cornell Hotel and Restaurant Administration Quarterly*, *43*(1), 39–46. https://doi.org/10.1016/S0010-8804(02)80007-6

Bastiaansen, M., Lub, X. D., Mitas, O., Jung, T. H., Ascencao, M. P., Han, D., Moilanen, T., Smit, B., & Strijbosh, W. (2018). Emotions as core building blocks of an experience. *International Journal of Contemporary Hospitality Management*, *31*(2), 651–668. https://doi.org/10.1108/IJCHM-11-2017-0761

Baudrillard, J. (1983). *Simulations*. Semiotext.

Belhassen, Y., Caton, K., & Stewart, W. P. (2008). The search for authenticity in the pilgrim experience. *Annals of Tourism Research*, *35*(3), 668–689. https://doi.org/10.1016/j.annals.2008.03.007

Bharwani, S., & Jauhari, V. (2013). An exploratory study of competencies required to co-create memorable customer experiences in the hospitality industry. *International Journal of Contemporary Hospitality Management*, *25*(6), 823–843. https://doi.org/10.1108/IJCHM-05-2012-0065

Bitner, M. J. (1992). Servicescapes: The impact of physical surroundings on customers and employees. *Journal of Marketing*, *56*(2), 57–71. https://doi.org/10.1177/002224299205600205

Bloemer, J., & de Ruyter, K. (1999). Customer loyalty in high and low involvement settings: The moderating impact of positive emotions. *Journal of Marketing Management*, *15*(4), 315–330. https://doi.org/10.1362/026725799784870388

Bolton, R. N., Gustafsson, A., McColl-Kennedy, J., Sirianni, N. J., & Tse, D. K. (2014). Small details that make a big difference: A radical approach to consumption experience as a firm's differentiating strategy. *Journal of Service Marketing*, *25*, 253–274. https://doi.org/10.1108/JOSM-01-2014-0034

Brown, L. (2013). Tourism: A catalyst for existential authenticity. *Annals of Tourism Research*, *40*, 176–190. https://doi.org/10.1016/j.annals.2012.08.004

Bruner, E. M. (1994). Abraham Lincoln as authentic reproduction: A critique of postmodernism. *American Anthropologist*, *96*(2), 397–415. https://doi.org/10.1525/aa.1994.96.2.02a00070

Bushell, R., & Sheldon, P. J. (2009). *Wellness and tourism: Mind, body, spirit and place*. Cognizant Communication.

Buxton, L. (2018). Destination spas and the creation of memorable guest experiences. *International Journal of Spa and Wellness*, *1*(2), 133–138. https://doi.org/10.1080/24721735.2018.1493778

Cannas, R. (2018). Diverse economies of collective value co-creation: The open monument event. *Tourism Planning & Development*, *15*(5), 535–550. https://doi.org/10.1080/21568316.2018.1505651

Checa-Gismero, P. (2018). Global contemporary art tourism: Engaging with Cuban authenticity through the Bienal de La Habana. *Tourism Planning & Development*, *15*(3), 313–328. https://doi.org/10.1080/21568316.2017.1399435

Clendenen, V. I., Herman, P. C., & Polivy, J. (1994). Social facilitation of eating among friends and strangers. *Appetite*, *23*(1), 1–13. https://doi.org/10.1006/appe.1994.1030

Cohen, E. (1979). A phenomenology of tourist experiences. *Sociology*, *13*(2), 179–201. https://doi.org/10.1177/003803857901300203

Cohen, E. (2007). Authenticity in tourism studies: Apres la Lutte. *Tourism Recreation Research*, *32*(2), 75–82. https://doi.org/10.1080/02508281.2007.11081279

Corsaro, D. (2020). Value co-destruction and its effect on value appropriation. *Journal of Marketing Management*, *36*(1–2), 100–127. https://doi.org/10.1080/0267257X.2019.1696876

Decre, B. G., & Pras, B. (2013). Simulating in-store lighting and temperature with visual aids: Methodological propositions and SOR effects. *The International Journal of Retail, Distribution and Consumer Research*, *23*(4), 363–393. https://doi.org/10.1080/09593969.2013.781050

Global Wellness Institute. (2018, October). *Global wellness economy monitor.*

Hanks, L., & Line, N. D. (2018). The restaurant social servicescape: Establishing a nomological framework. *International Journal of Hospitality Management*, *74*, 13–21. https://doi.org/10.1016/j.ijhm.2018.01.022

Harris, L. C., & Ezeh, C. (2008). Servicescape and loyalty intensions: An empirical investigation. *European Journal of Marketing*, *42*(3–4), 390–422. https://doi.org/10.1108/03090560810852995

Hemmington, N. (2007). From service to experience: Understanding and defining the hospitality business. *The Service Industries Journal*, *27*(6), 747–755. https://doi.org/10.1080/02642060701453221

Holbrook, M. B. (1996). Customer value—a framework for analysis and research. *Advances in Consumer Research*, *23*, 138–141.

Huang, Y.-C., Chen, C.-C., & Gao, M. J. (2019). Customer experience, well-being, and loyalty in the spa hotel context: Integrating the top-down & bottom -up theories of well-being. *Journal of Travel and Tourism Marketing*, *36*(5), 595–611. https://doi.org/10.1080/10548408.2019.1604293

Javri, H., Kerenen, J., Ritala, P., & Vilko, J. (2020). Value co-destruction in hotel services: Exploring the misalignment of cognitive scripts among customers and providers. *Tourism Management*, *77*, 1–13. https://doi.org/10.1016/j.tourman.2019.104030

Kim, K. T., Bae, J., Kim, J.-C., & Lee, S. (2016). The servicescape in the fitness center: Measuring fitness center's service. *International Journal of Sport Management, Recreation and Tourism*, *21*, 1–20. http://dx.doi.org/10.5199/ijsmart-1791-874X-21a

Kitchen, J. (2019). Top team: Rosewood. *Spa Business 3*, 16–17.

Krishna, A., Lwin, M. O., & Morrin, M. (2010). Product scent and memory. *Journal of Consumer Research*, *37*(1), 57–66. https://doi.org/10.1086/649909

Kucukusta, D., & Guillet, D. B. (2014). Measuring spa-goers' preferences: A conjoint analysis approach. *International Journal of Hospitality Management*, *41*, 115–124. https://doi.org/10.1016/j.ijhm.2014.05.008

Laing, J., Voigt, C., & Frost, W. (2014). Fantasy, authenticity and the spa tourism experience. In C. Voigt & C. Pforr (Eds.), *Wellness tourism: A destination perspective* (pp. 220–233). Routledge.

Le, T. H., Arcodia, C., Novais, M. A., & Kralj, A. (2019). What we know and do not know about authenticity in dining experiences: A systematic literature review. *Tourism Management*, *74*, 258–275. https://doi.org/10.1016/j.tourman.2019.02.012

Lin, I. Y., & Mattila, A. S. (2018). A conceptual model of co-creating an authentic luxury spa experience. *International Journal of Spa and Wellness Management*, *1*(1), 39–54. https://doi.org/10.1080/24721735.2018.1438537

Loureiro, S. M. C., Almeida, M., & Rita, P. (2013). The effect of atmospheric cues and involvement on pleasure and relaxation: The spa hotel context. *International Journal of Hospitality Management*, *35*, 35–43. https://doi.org/10.1016/j.ijhm.2013.04.011

Lovelock, C. H. (1996). *Services marketing* (3rd ed.). Prentice-Hall.

MacCannell, D. (1973). Staged authenticity: Arrangements of social space in tourist settings. *The American Journal of Sociology*, *79*(3), 589–603. https://doi.org/10.1086/225585

Mari, M., & Poggesi, S. (2011). Servicescape cues and customer behavior: A systematic literature review and research agenda. *The Service Industries Journal*, *33*(2), 171–199. https://doi.org/10.1080/02642069.2011.613934

McIntosh, A. J., & Siggs, A. (2005). An exploration of the experiential nature of boutique accommodation. *Journal of Travel Research*, *44*(1), 74–81. https://doi.org/10.1177/0047287505276593

Megson, K. (2019). Jungle VIP. *Spa Business 2*, 70–74.

Mehmetoglu, M., & Engen, M. (2011). Pine and Gilmore's concept of experience economy and its dimensions: An empirical examination in tourism. *Journal of Quality Assurance in Hospitality and Tourism*, *12*(4), 237–255. https://doi.org/10.1080/1528008X.2011.541847

Michopoulou, E. (2017). Marketing for the spa industry. In S. Rawlinson & T. Heap (Eds.), *International spa management* (pp. 114–130). Goodfellow Publishers.

Milliman, R. E. (1986). The influence of background music on the behaviour of restaurant patrons. *Journal of Consumer Research*, *13*(2), 286–289. https://doi.org/10.1086/209068

Mody, M., Hanks, L., & Dogru, T. (2019). Parallel pathways to brand loyalty: Mapping the consequences of authentic consumption experiences for hotels and Airbnb. *Tourism Management*, *74*, 65–80. https://doi.org/10.1016/j.tourman.2019.02.013

Nakata, R., & Kawai, N. (2017). The "social" facilitation of eating without the presence of others: Self-reflection on eating makes food taste better and people eat more. *Physiology & Behaviour*, *179*, 23–39. https://doi.org/10.1016/j.physbeh.2017.05.022

Newman, G. E., & Smith, R. K. (2016). Kinds of authenticity. *Philosophies Compass*, *11*(10), 609–618. https://doi.org/10.1111/phc3.12343

Nilsson, E., & Ballantyne, D. (2014). Re-examining the place of servicescape in marketing: A service-dominant logic perspective. *Journal of Services Marketing*, *28*(5), 374–379. https://doi.org/10.1108/JSM-01-2013-0004

Oh, H., Fiore, A. M., & Jeoung, M. (2007). Measuring experience economy concepts: Tourism applications. *Journal of Travel Research*, *46*(2), 119–132. https://doi.org/10.1177/0047287507304039

Öznalbant, E., & Alvarez, M. D. (2020). A socio-cultural perspective on Yoga tourism. *Tourism Planning & Development*, *17*(3), 260–274. https://doi.org/10.1080/21568316.2019.1606854

Pine, B. J., & Gilmore, J. H. (1998, July–August). Welcome to the experience economy. *Harvard Business Review*, *76*(4): 97–105.

Pine, B. J., & Gilmore, J. H. (1999). *The experience economy: Work is theatre & every business a stage*. Harvard Business School Press.

Pine, J. B., & Gilmore, J. H. (2011). *The experience economy*. Harvard Business School Publishing.

Pizam, A. (2010). Creating memorable experiences. *International Journal of Hospitality Management*, *29*(3), 343. https://doi.org/10.1016/j.ijhm.2010.04.003

Ple, L., & Caceres, R. C. (2010). Not always value co-creation: Introducing interactional co-destruction of value in service-dominant logic. *Journal of Services Marketing*, *24*(6), 420–437. https://doi.org/10.1108/08876041011072546

Poluzzi, I., & Esposito, S. (2020). Implications of rituals and authenticity within the spa industry. *International Journal of Spa and Wellness*, *2*(2), 98–106. https://doi.org/10.1080/24721735.2020.1770984

Pullman, M. E., & Gross, M. A. (2004). Ability of experience design elements to elicit emotions and loyalty behaviors. *Decisions Sciences*, *35*(3), 551–557. https://doi.org/10.1111/j.0011-7315.2004.02611.x

Reisinger, Y., & Steiner, C. J. (2006). Reconceptualizing object authenticity. *Annals of Tourism Research*, *33*(1), 65–86. https://doi.org/10.1016/j.annals.2005.04.003

Reitsamer, B. F. (2015). Post-consumptive experience in servicescapes: The impact of mental re-enactment on consumers' loyalty. In *Proceedings of the AMA Winter Educators' Conference* (pp. 6–14). American Marketing Association.

Rosenbaum, M. S., & Massiah, C. (2011). An expanded servicescape perspective. *Journal of Service Management*, *22*(4), 471–490. https://doi.org/10.1108/09564231111155088

Sanchez-Fernandez, R., & Iniesta-Bonillo, M. A. (2007). The concept of perceived value: A systematic review of the research. *Marketing Theory*, *7*(4), 427–451. https://doi.org/10.1177/1470593107083165

Sharpley, R. (2018). *Tourism, tourists and society* (5th ed.). Routledge.

Sipe, L. J., & Testa, M. R. (2018). From satisfied to memorable: An empirical study of service and experience dimensions on guest outcomes in the hospitality industry. *Journal of Hospitality Marketing and Management*, *27*(2), 178–195. https://doi.org/10.1080/19368623.2017.1306820

Skalen, P., Gummerus, J., von Koskull, C., & Magnusson, P. R. (2015). Exploring value propositions and service innovation: A service- dominate logic study. *Journal of the Academy of Marketing Science*, *43*(2), 137–158. https://doi.org/10.1007/s11747-013-0365-2

Smith, J. B., & Colgate, M. (2007). Customer value creation: A practical framework. *Journal of Marketing Theory and Practice*, *15*(1), 7–23. https://doi.org/10.2753/MTP1069-6679150101

Steiner, C. J., & Reisinger, Y. (2006). Understanding existential authenticity. *Annals of Tourism Research*, *32*(2), 299–316. https://doi.org/10.1016/j.annals.2005.08.002

Tabacchi, M. H. (2010). Current research and events in the spa industry. *Cornell Hospitality Quarterly*, *51*(1), 102–117. https://doi.org/10.1177/1938965509356684

Thorne, F. C. (1963). The clinical use of peak and nadir experience reports. *Journal of Clinical Psychology*, *19*(2), 248–250. https://doi.org/10.1002/1097-4679(196304)19:2<248::AID-JCLP2270190236>3.0.CO;2-D

Tiberghien, G. (2019). Managing the planning and development of authentic eco-cultural tourism in Kazakhstan. *Tourism Planning & Development*, *16*(5), 494–513. https://doi.org/10.1080/21568316.2018.1501733

Tombs, A., & McColl-Kennedy, J. R. (2003). Social-servicescape conceptual model. *Marketing Theory*, *3*(4), 447–475. https://doi.org/10.1177/1470593103040785

Turley, L. W., & Milliman, R. E. (2000). Atmospheric effects on shopping behavior: A review of the experimental evidence. *Journal of Business Research*, *49*(2), 193–211. https://doi.org/10.1016/S0148-2963(99)00010-7

Vargo, S. L., & Lusch, R. F. (2004). Evolving to a new dominant logic for marketing. *Journal of Marketing*, *68*(1), 1–17. https://doi.org/10.1509/jmkg.68.1.1.24036

Vargo, S. L., & Lusch, R. F. (2008). Service-dominant logic: Continuing the evolution. *Journal of the Academy of Marketing*, *36*(1), 1–10. https://doi.org/10.1007/s11747-007-0069-6

Vargo, S. L., & Lusch, R. F. (2017). Service-dominant logic 2025. *International Journal of Research in Marketing*, *34*(1), 46–67. https://doi.org/10.1016/j.ijresmar.2016.11.001

Walls, A., Okumus, F., Wang, Y., & Joon-Wuk Kwun, D. (2011). Understanding the customer experience: An exploratory study of luxury hotels. *Journal of Hospitality Marketing and Management*, *20*(2), 166–197. https://doi.org/10.1080/19368623.2011.536074

Wang, N. (1999). Rethinking authenticity in tourism experience. *Annals of Tourism Research*, *26*(2), 349–370. https://doi.org/10.1016/S0160-7383(98)00103-0

Wuttke, M., & Cohen, M. (2008). Spa retail. In M. Cohen & G. Bodeker (Eds.), *Understanding the global spa industry: Spa management* (pp. 208–220). Butterworth-Heinemann.

Yang, L. (2019). Cultural tourism in a replicated old town: Tourists' views. *Tourism Planning & Development*, *16*(1), 93–111. https://doi.org/10.1080/21568316.2018.1470998

Yang, S., & Namkung, Y. (2009). Perceived quality, emotions, and behavioural intentions: Application of an extended Mehrabian—Russell model to restaurants. *Journal of Business Research*, *62*(10), 451–460. https://doi.org/10.1016/j.jbusres.2008.05.006

Zatori, A., Smith, M., & Puczko, L. (2018). Experience-involvement, memorability and authenticity: The service provider's effect on tourist experiences. *Tourism Management*, *67*, 111–126. https://doi.org/10.1016/j.tourman.2017.12.013

Ziethanl, V. A. (1988). Consumer perceptions of price, quality and value: A means-end model and synthesis of evidence. *Journal of Marketing*, *52*(3), 2–22. https://doi.org/10.1177/002224298805200302

Destination Image Co-creation in Times of Sustained Crisis

Kyriaki Glyptou

ABSTRACT
Customer co-creation feeds from customer engagement, value recognition and experience appreciation. Tourists participation in the image communication of a destination in adversity is well documented along literature addressing their motivations and their reliability as information intelligence. What remains still vague is an exploration of the above dynamics in the case of destinations in sustained crisis hence the customer predispositions for destinations under an extended duration yet reduced intensity turbulent destination image. Using Lesvos (Greece) as a case study of a destination affected by refugee and immigrant mobilities since 2012, this paper explores those constructs affecting tourists' response and engagement in the formulation, promotion and hence co-creation of an affected destinations' cognitive and affective image. The theoretical contribution of the paper lies in the exploration of the conscious and unconscious tourist triggers that could promote the co-repair and co-restoration of a long-affected destination' image, with direct managerial implications both for destination and crisis management.

Introduction

Destination management and branding literature defines a destination's image along the set of beliefs, expectations and emotional thoughts tourists create of a place (Kotler et al., 2017; Zhang et al., 2014). Destination images are inherently perceptual subject to the personality attributes (psychographics) of the tourist, but also strongly affected by the underlying circumstances of the visiting destination. Soenmez and Graefe's (1998) travel framework suggests that regardless of the personality characteristics, perceptions of destination risk and safety are amongst the key antecedents of travel intentions. The latest acquires particular interest for destinations in crisis, with an increasing body of literature exploring the image recovery implications after short-term crisis events such as natural disasters or terrorist attacks (e.g. Bauer, 1960; Coombs, 2015; Kozak et al., 2007; Nasir & Yilmaz, 2017). Yet, similar implications for destinations' in sustained crisis, like the ongoing mobility of refugees and immigrants through Mediterranean tourist destinations, have only been minimally addressed. After the first repercussions of the crisis outbreak on the image of affected destinations almost seven years ago, and with reduced media attention since then despite the continuation of the phenomenon, what remains to be explored is the process of image co-creation for destinations perceived to be in sustained crisis.

Perceptions over the refugee and immigrant mobility carries inherently a strong personal bias. The terms are mistakenly used interchangeably when, as summarised by Farmaki and Christou (2019, p. 671), their distinction lies in one's "impediment to safely return to their home country", which is however not always a clear line. In times of anthropogenic crises, such as those related to any populations' mobility, the international community engages directly or indirectly both in the generation and/or circulation of opinion statements and news (Tucker, 2016). Online information sharing platforms and social media facilitate the generation speed and volume of user-generated content (UGC), thus serving simultaneously as opinion sharing channels and information intelligence media (Wang et al., 2016; Williams et al., 2017), triggering consciously and unconsciously potential tourists' image perceptions and hence, visiting intentions for an undefined period of time.

The research develops on the island of Lesvos, the biggest island of the North Aegean region in Greece with a permanent population of 86,436 inhabitants according to the 2011 census. Lesvos is not a tourism destination in the traditional term of the context, nor has a distinctive tourism destination brand. It is a destination with loyal repeaters of an older age, and with a high number of second-home owners who often have a strong emotional affiliation and relation to the island. Refugee and immigrant inflows have been first recorded since 2012, yet Lesvos reached its pick in 2015 with a number of 271,156 arrivals only between January to August of the year (Tsartas et al., 2019). With the numbers subsiding over the years, the island currently and regularly hosts approximately 23,000 of immigrants in hotspots (Deutsche Welle, 2020) which correspond to a quarter of its population. This ongoing situation has generated over the years, frustration both from locals and immigrants and is often exhausted in demonstrations and riots (Deutsche Welle, 2020).

A number of authors have approached the particularities of the refugee crisis in the islands of the North Aegean emphasising on the supply perceptions of either the service industries (Farmaki & Christou, 2019; Pappas & Papatheodorou, 2017) or other local stakeholders (Tsartas et al., 2019). Demand-side approaches on destination image and visiting intentions were also performed, yet they either addressed implication at national level (Zenker et al., 2019) or indirect effects towards alternative destinations (Cirer-Costa, 2017). This literature addressed issues primarily during the pick of the response phase of the refugee crisis. With ongoing inflows of reduced intensity over a prolonged period of time, and with an inconsistent media and exposure attention, it is still necessary to explore the image implications and the image recovery process for destination consumers. In order to contribute towards the destination image formulation and recovery body of literature, the aim of the current research is to explore the conscious and unconscious tourists' response and engagement in the co-creation of the image of a destination in sustained crisis.

In acknowledging the contribution of tourists as active creators rather than passive recipients of information, the research adopts the service-dominant logic (Lusch & Vargo, 2006) to conceptualise customer co-creation along the participatory process of formulating and evolving the core offering and experience itself. When focusing on a destination's image, and in line with Tasci and Gartner (2007), the process is extended to consider customer's active response and engagement in the cumulative interpretation of meanings, feelings and behaviours associated with a destination. Considering the sensitivity and the ethical limitations of the topic, a quantitative approach was deemed most appropriate to explore the theoretical constructs dictating tourists' predispositions and behaviour and holistically capture their implications for the destination image co-creation (Lee, 1993).

Building on the particularities of Lesvos as case study destination, this research aims to contribute to the growing body of literature on the implications of global population mobilities and the organic crisis response mechanisms at tourism destination level. Its theoretical contribution lies in the broader field of destination and crisis management, through the exploration of the conscious and unconscious tourist contribution to the image repair and restoration in destinations in sustained crisis. More specifically, the paper explores those constructs triggering tourists' engagement in the formulation, promotion and hence co-creation of the affected destinations' cognitive and affective image. From a managerial perspective, the paper contributes to the enhancement of destinations' resilience through the identification of catalysts of tourists' behaviour and the appreciation of their role in the recovery of its image.

Theoretical constructs and research hypotheses

Refugee crisis perceptions

In times of crisis, refugees and immigrants, along national and international designated support (Red Cross, IOM) and control (Frontex, troop forces) groups could turn amongst the most prominent groups of people at destinations. According to the Group Threat Theory (Atwell Seate, & Mastro, 2015; Blumer, 1958; Croucher, 2013; Stephan, 2014) members of the dominant locals ingroup might feel threatened by an outgroup, such as refugees and immigrants, when they cognitively or perceptually consider them invasive or threatening for their collective or individual interests including safety, security, resources and overall quality of life. An exploration of their effect on the image and brand of the affected destination, or the motivations of tourists to visit while and after the crisis remains limited and only gets attention in light of the current mobilities in the Mediterranean region (Seetaram, 2012; Simpson et al., 2016).

The predispositions and willingness to visit a destination comes down to the three major antecedents of the travel destination choice process, namely: external, internal and demographic factors (Sönmez & Graefe, 1998) with the first two acquiring particular importance in the context of destinations in crisis. The success of a tourism destination depends greatly on its ability to provide a safe and secure environment for visitors (Stylidis et al., 2017), with Khan et al. (2017) suggesting that even isolated incidents may have a strong effect on tourists' perceptions. When it comes to internal predictors of travel intentions though, the perceptual image of safety in a destination is muddled by the personal set of beliefs, attitudes, expectations and previous experiences (Baloglu & McCleary, 1999; Tan & Wu, 2016).

Perceptions over the refugee and immigrant mobility carries inherently a strong bias from one's personal beliefs and attitudes. Other than one individual's psychographic profile (origin, age, gender, income, level of education), Simpson et al. (2016) suggest that the political ideology and orientation is very likely to influence visiting intention and destination image both due to the attitudes towards the refugees and immigrants, but also towards the security and control forces. Political orientation in the context of social and political issues extends in the spectrum from conservatism to liberalism

(Bierbrauer & Klinger, 2002). In line with the Group Threat Theory, visitor political orientation in the context of a destination affected by refugee and immigrant mobility relates to perceptions of altruism or prejudice against outgroups, their correlation to risks and security issues at destination level (Fuchs et al., 2013), or even the preference for visiting destinations congruent with their own political ideology expressed on the way they treat outgroups (Legg et al., 2012). In order to further explore the relationship between personal beliefs and the perceptual image of a destination affected by outgroup mobilities, the paper proposed the following research hypothesis:

> *H1:* Perceptions over the refugee/immigrant crisis will have a significant effect on the destination image prior to visitation.

Marketing influence

Other than pure personal interest push factors, an individual's intention to visit a destination is traditionally triggered by promotional material, opinions of travel agents, the media, popular culture, the word of mouth (WOM) from family and friends (Rodríguez Molina et al., 2013) or increasingly the electronic word of mouth (eWOM). The messages behind the identifiable and anonymous promotional and branding activities conjure up a destination image that influences tourists' cognitive expectations and their final travelling intention. Marketing influence has thus a crucial role in comforting potential customers and reducing any uncertainty and risks associated with their visiting intention, as tourists would prefer destinations that aspire trust (Kock et al., 2016; Zenker et al., 2019).

Information intelligence is particularly important for tourists intending to visit destinations in crisis. Simpson et al. (2016) identify information from mass and social media as key image catalysts for destinations exposed to the frequency and intensity of crisis stories. Other than the broadly recognised speed and pluralism of circulating and openly commenting on information, research findings identify two variant trends. On one hand, official identifiable sources are perceived of higher credibility and hence a greater impact on public opinion (Takahashi et al., 2015; Tasci & Gartner, 2007), while on the other, anonymous eWOM appears also as highly influencial due to its perceived experiential nature and lack of underlying incentives (Duverger, 2013; Filieri, 2015). In any of the cases, the post service and experience information intelligence are considered more reliable to reflect the actual exchange value, with the latest considered as a very accurate predictor of purchasing intentions (Chen & Chen, 2010).

Regardless of the source of information, crisis stories tend to spread faster and likely trigger generic or destination specific WOM which consciously and unconsciously affects potential tourists' expectations, image perceptions and hence, visiting intentions (Matzler et al., 2016). The overdramatisation of information from the visitor side, often bears fears, anticipations and past experiences of their own and surrounding environment leading to perceptual rather than cognitive evaluations of the destination image (Simpson et al., 2016) which are often in the sphere of generic rather than destination-specific assumptions. It is then the strength of the reputation and the trust on the destination brand that could reverse negative connotations and potential damage. Considering the multiple marketing intelligence channels and their implications on the formulation of a destination image, the paper proposes the following research hypothesis:

H2: Marketing influences will have a significant effect on the destination image prior to visitation.

Destination image

For a tourist, a destination's image is the amalgamation of subjective perceptions of reality and combination of pieces of information (Kotler et al., 1993; Qu et al., 2011). Even if destination trust and loyalty under regular circumstances are leading concepts on destination image and branding research, little has been written on their interrelationship and their ability to predict tourists' behaviours for destinations in crisis (Zenker et al., 2019). The point of interest is thus, whether the strength of the brand or previous image formulation of the destination is strong enough to overcome any cognitive or perceptual fear, risk and uncertainty of the impacted destination, hence proceed with the purchasing intention.

Research on destination image has been grounded on Gartner's (1994) cognitive, affective, and conative attribute typologies, which led most research on re-visiting intentions under regular destination circumstances (Chew & Jahari, 2014; Qu et al., 2011). In the context of destinations in crisis, each appraisal gets a more expansive meaning. More specifically, the cognitive image component relates to attitudes and informed evaluations of a destination's level of safety, security and ability to deliver to the expected standards, while the affective image bears a more emotional response to the underlying circumstances directly linked to the psychographic profile of the potential tourist. Both cognitive and affective components have a strong influence on the pre-visit destination image of a destination and the expected experience (Stylos et al., 2017). Finally, the conative image component builds on the personal evaluation of the pre-visit destination image to actively consider the destination as travelling potential (White, 2014). This active consideration might be manifested both through the actual purchasing behaviour or even through the communication and engagement on positive or negative image content creation about the actual destination.

The differentiation and conceptualisation of the various image components for destinations in crisis is beyond the scope of this research. Recognising the challenges for their operationalisation though (Stylos et al., 2017), the focus of this current research is to explore their combined influence on the attitudinal behaviours of tourists either before or during their visit on the affected destination to actively engage in the co-creation of the overall (holistic) destination image. In fact, this need for a combined integrative approach becomes essential when needing to predict tourists' attitudinal outcomes for destinations in crisis due to the subtle differentiations in the mind of the potential customer. To explore the above dynamics further, this research proposes the following hypotheses:

H3: Pre-visit destination image will have a significant effect on the co-creation of the destination image.

H4: Pre-visit destination image will have a significant effect on the current (during the visit) destination image.

H5: Current (during the visit) destination image will have a significant effect on the co-creation of the destination image.

User-generated content (UGC)

Over the last years, social media and information sharing platforms have boosted the synchronous and asynchronous distribution of UGC, turning eWOM as a key influencer for destination perceptions and travel decision-making processes (Hudson & Thal, 2013; Williams et al., 2017). Tourist generated content, particularly while at destination, is perceived as more authentic and credible to encompass the multiple perspectives of a destination brand or tourism experience (Lamsfus et al., 2014; Tussyadiah, 2015). Visiting a destination in crisis, can trigger UGC particularly in terms of the feeling of safety, security and perceptions of threat. Other than observing from the distance actual events, followers are primarily interested in the experiential communication of safety and security information from other people they could easily relate to or identify with. In that regard, tourist contribution in communicating information on the actual happenings and expressing solidarity to the involved stakeholders, even other tourists themselves, maintains a key role in the formulation of the destination's image and facilitate the recovery of its reputation (Ulmer et al., 2007; Ye et al., 2017).

The content generated by tourists is not just factual, but mainly carries emotional and moral predispositions towards the involved stakeholders and destination (Mair et al., 2016). In the case of refugee mobilities the sentiments of solidarity and empathy could be addressed either towards the refugees, or the affected locals or even towards both. Crisis communication is key in destination image recovery after a period of time though and in cases of sustained crisis, tourist UGC rarely entails the element of event surprise anymore (Ulmer et al., 2007). Instead, it reflects more the actual experience at the destination and any other element of the trip that might confirm or contradict their initial expectations. Another common practice amongst tourists is to follow and be-friend the official and unofficial channels of destination promotion (NTOs, DMOs) to keep up to date with tourist activities and events during their stay. It is often the case, that they intentionally or unintentionally show empathy towards the published information and engage in the promotion of the destination through the reaction and attraction (sharing and liking) to relevant content (Usakli et al., 2017). Higher numbers of solidarity messages and destination promotion postings generate reactions of empathy and contribute more effectively in the image recovery of destinations in crisis (Oliveira & Huertas, 2019). Considering the dynamic generation of content through social media and information sharing platforms, as well as the enabling of real-time tourist-followers interaction, this research proposes the following hypothesis:

> *H6:* UGC generated during the visit will have a significant effect on the co-creation of the destination image.

Tourist interactions at destinations

Building on Plog's (2016) traveller profile spectrum, tourist behaviour ranges along a continuum from dependable psychocentric to venturer allocentric travellers. Their differentiation revolves around various personality traits that manifest through perceptions and responses to risk, threat, and comfort zone, hence affecting the overall perception of a destination's image. Even if mainly discussed in the context of destinations under "normal" circumstances, an exploration of how they relate to destinations in sustained crisis is still limited. Literature suggests that tourists encounters in a destination will relate to both the local population (the tourist–host interaction) and other tourists (tourist-to-tourist) interaction (Huang & Hsu, 2010; Pearce, 2005). Within the context of this research, interactions will include a third group, the outgroup of refugees and immigrants. All interactions are discussed in the context of their individual and aggregated impact in shaping the image of the host destination.

Tourist-to-tourist interactions might be either direct interpersonal in the context of shared activities, events, or service providers or indirect in the broader destination environment all contributing to the co-creation of the trip experience and hence the destination image (Loi & Pearce, 2012). Other than the quality of those encounters (generated feelings of enjoyment or frustration), Yang (2016) identifies their intensity and frequency to equally influence the destination image formation. The impact of tourist-to-tourist interactions has been generally discussed in the context of intense incidents or subtle encounters in destinations under "normal" circumstances, yet little has been written in the context of destinations in crisis.

Tourists interact with local hosts either directly through the consumption of tourism services or in the context of their broader encounters. Regardless of the characteristics of the destination or the product, host–guest interactions are amongst the most frequent and authentic ones, hence dictating a great deal of the destination experience. Positive experience feedback on host–guest interactions induces a higher level of affective attachment, cohesion and empathy which link directly to an affective image of a destination and favourable WOM intentions or repeat visit tendencies (Xu et al., 2019). On the other hand, frustration or the general feeling of distress and unrest from local's end, might make tourists feel uncomfortable and unwelcome in a destination. This often does not come as a surprise for destination in sustained crisis, where the ongoing societal turbulence might have impacted the wellbeing and resilience of the local hosts and their ability to maintain the standards of hospitality. For crises related to ingroup mobility, it might be possible for tourists to be treated aggressively or unfairly in public places, since they might be confused with members of the ingroup (refugees or immigrants).

Group threat theory suggests that ingroup mobility distracts multiple aspects of the destination balance. Ingroups are often associated with criminal or malcontent activity which puts in jeopardy the feeling of safety and security at destinations and generates the feeling of unrest and distress (Simpson et al., 2016; Sönmez et al., 2016). Tourists may experience this either indirectly through their interactions with hosts in the broader tourism experience environment or through their direct encounters with the outgroup. Depending again on the quality and the intensity of these interactions and their psychographic characteristics, tourist perceptions might range in a continuum from solidarity to fear, both reduced to their perceived safety and level control at destination level (Woosnam et al., 2015). This triggers the intentional or unintentional generation of destination content and WOM which only communicates their experience at the destination. What is interesting in the case of ingroups and tourist interactions though, is that they could potentially affect the whole service experience and process, as tourists could potentially envision on them the direct threat and cause of their anxiety (Stephan, 2014). Cirer-Costa (2017) discussed how the direct experience of tourists with immigrant camps in Mediterranean destinations during their pick of the crisis, has affected their image of a destination and impacted their predispositions to return and recommend the area to others.

All the above interim encounters constitute components of the overall interactions at destination level. Research on their specific delineation and interlinkages of their underlying dynamics in destinations in crisis is increasing, as are their implications for the destination image. Still what remains interesting to explore is the duration and intensity of that impact, meaning the amount of time that image stays with the tourists and the amount of time it takes to recover, which obtains particular importance for destinations in sustained crisis. Building on crisis management literature (Pappas, 2019; Paraskevas & Arendell, 2007) consistent law enforcement initiatives and the attainment of a feeling of stability and control is what primarily recovers image perceptions in crisis destinations. Still as main catalyst remains the psychographic profile of the tourists and its interpretation of happening and experiences while at destination. To further explore the dynamics of interactions in destinations in sustained crisis as well as their impact on the co-creation of its image, this research proposes the following hypotheses:

H7: Tourist to tourists' interactions during the visit will have a significant effect on the co-creation of the destination image.

H8: Host to tourists' interactions during the visit will have a significant effect on the co-creation of the destination image.

H9: Refugees to tourists' interactions during the visit will have a significant effect on the co-creation of the destination image.

H10: Overall tourist interactions during the visit will have a significant effect on the co-creation of the destination image.

Destination image co-creation

On the basis of experience economy, research and practice have acknowledged the importance for the tourist consumer to actively participate and contribute in the delivery of the purchased experience, hence co-create its quality, value and meaning (Pine & Gilmore, 1998; Suntikul & Jachna, 2016; Waligo, 2013). In contrast to a destination's brand which reflects its attributes from the provider's perspective (Kapferer, 2008), a destination image is formulated in the mind of an actual and potential customer and resides in its perceived or expected experience value and satisfaction at the destination. Whether cognitive, affective or conative, a destination image is strongly associated with the cumulative interpretation of perceived meaning, hence, a destination's image co-creation relates to a tourist's active engagement and internalisation process of the perceived exchange value created and quality of experience generated from visiting the destination.

The implications of perceived safety and threat on a destination's image are well-documented in the international literature (George, 2010; Yüksel & Yüksel, 2007), as is tourists' expectation to be treated with respect and hospitality. The more authentic and engaging their experience, the higher tourists' perception of exchange value creation at the destination (Chen & Tsai, 2007; Choo & Petrick, 2014). Regardless of the behaviour and psychographic traits that influence their co-creation engagement while at the destination, what remains still unclear is their implications for destination commitment and loyalty, which translates into positive (e)WOM, repeat visit tendency and a positive affective image of the destination overall (Nasir & Yilmaz, 2017).

The conceptualisation and operationalisation of the attitude construct as a cognitive and affective perception of visiting intentions has been early documented by Soenmez and Graefe (1998). Yet, tourist predispositions towards destinations with specific underlying characteristics such as sustained crises have not been adequately addressed in the context of destination image co-creation. Ortiz et al. (2015) amongst others, suggests that tourists often alter their perceptions and attitudes of a destination image after acquiring additional organic information intelligence, which often changes their overall travel behaviour and intentions. Once the breakout of a crisis is over what is essential for destinations is to convince tourists to overcome the perceptual fear and inclination towards a destination and proceed with their visit in order to build personal exchanges and encounters, hence an evidence-based cognitive image. The antecedents of their behaviour and engagement remain subject to their experience at the destination.

Proposed destination image co-creation model

The theoretical grounds of the proposed model (Figure 1) build throughout on the Theory of Planned Behaviour (TPB) and Perceived Risk Theory (PRT) to suggest that tourism engagement in the image co-creation of destinations in sustained crisis is dependent on the formulation of the pre-visit and during visit destination image, the overall quality and intensity of interactions they had at destination, as well as their predispositions to generate and share content in the form of information intelligence. For the specifics of destinations affected by refugee and immigrant mobility, overall interactions are conceptualised along encounters with three primary groups: other tourists, local host and refugees, while the formulation of pre-visit destination image is considered to be influenced both by tourists' personal beliefs and attitudes on refugee mobility as well as the impact of marketing and promotion to their purchasing behaviour and visiting intentions.

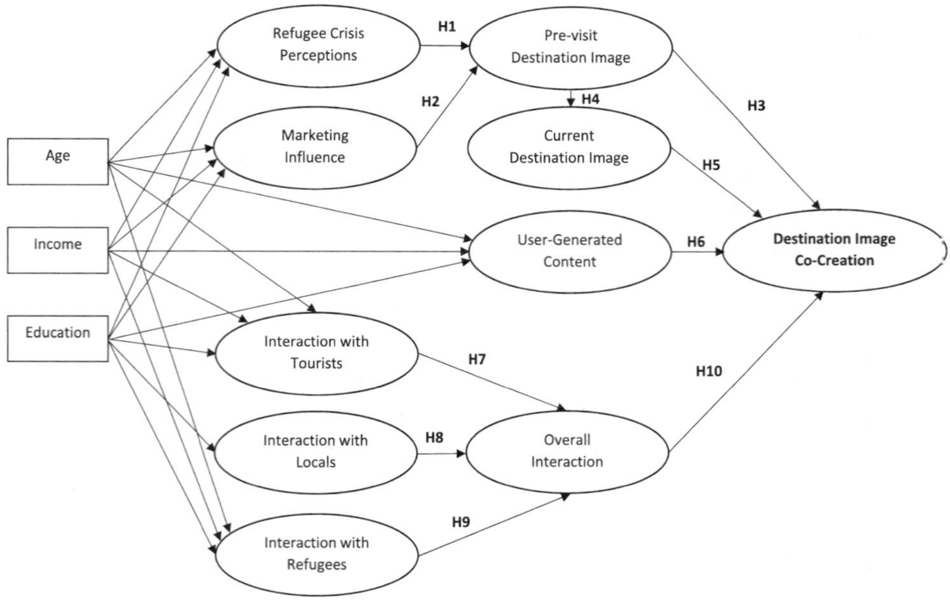

Figure 1. Proposed destination image co-creation model.

TPB supports the identification of purchasing behaviours (here visiting intentions of a destination is crisis) based on a set of perceptual assumptions that dictate tourists' motivations and intentions (Ajzen, 1991). In an era of technological advancement and high dependency on e-image and online competition, TPB is widely used to explain and predict consumer purchasing intentions through the evaluation of the underlying factors influencing customer decision-making (Pappas, 2016). More importantly, TPB has been applied in the study of tourist purchasing intentions under uncertainty (Quintal et al., 2010), hence it is considered appropriate to support the exploration of tourism behaviour, intentions and perceptions formulation for destinations in crisis. Similarly, PRT suggests that risk aversion is relevant to the perceived loss, with customers tending to develop risk reduction mechanisms to cope with decision-making under uncertainty (Bauer, 1960). Enhancing information intelligence as an uncertainty and risk reduction strategy is well documented in the international literature of many academic and managerial disciplines. In the current digital era, what remains still under discussion is the impact of the plethora of available information, the speed of content generation and sharing and the credibility of the source of information. In the context of destinations in crisis, Kaplan et al. (1974) components of tourism intentions and perception of risk are adapted to cover primarily safety (physical) risk and product value (product performance), while the financial (price) and privacy-time loss risks are not deemed so relevant.

Research method

Participants

The research was conducted on a sample of 550 domestic and international tourists who visited the island of Lesvos, North Aegean Region of Greece during August and September 2019. Considering the sensitivity and the ethical limitations of the issues in question, a quantitative approach administered through a closed structured questionnaire was deemed as most appropriate tool to collect anonymous data on sensitive issues without compromising on the response rate (Lee & Sargeant, 2011). Face to face questionnaire collection was preferred to the online version in order to capture tourists' behaviours and attitudes while still at destination. The questionnaire was administered in the English language. Participants were randomly approached in public spaces in Mytilene (the capital of the island) an area which has attracted a lot of media attention due to being affected by refugee and immigrant mobilities since 2012. Participants were selected merely on the basis of being on the island; no further age, digital literacy or familiarity to the destination exclusion criteria were applied in the sample. Missing data were excluded listwise to allow the smoother analysis of the sample; a total of 513 (93% response rate) completed questionnaires were used for the analysis.

Sample size

The sample size was calculated by means of Raosoft considering a population size (tourist arrivals in Lesvos) of over 80,000 tourists for the year 2018, which suggests an appropriate sample size of over 380 participants. The sample size recommendation doesn't change much for larger populations (Raosoft.com), leaving the 513 samples as more than sufficient to extract reliable conclusions on a <5% (4.32%) margin of error and a 95% confidence level. The proposed model fit was tested by means of the χ^2, while the components validity and reliability was tested through loadings and Cronbach's Alpha. Structural Equation Modelling (SEM) was then employed to explore the linearity of relationships amongst the studied multivariate constructs.

Measures

The questionnaire is based on nine key constructs delineated in 53 control statements. Other than three questions on participants age, income and level of education, the control statements were developed along a 1–5 Linkert scale (1: Strongly Agree; 2: Agree; 3:Neither Agree nor Disagree; 4:Disagree; 5: Strongly Disagree and 6: Not relevant/applicable in my case). Their validity and reliability abide by the recommendation of Gross and Brown (2008) . The statements were identified in seven previous studies; none of them has been reverse coded for the purposes of this study. More specifically, the statements on refugee and immigrant crisis perceptions were based on Simpson et al. (2016) while that on marketing influence on Pappas (2019). The formulation of pre-visit and current destination image was based on the work of Ramseook-Munhurrun et al. (2015). UGC tendencies were inspired by the work of Oliveira and Huertas (2019). Yang (2016) proposed a scale for the assessment of interactions amongst tourist; to enhance the consistency and coherence of the research the same scale of statements was applied to assess both the constructs of interactions with locals and refugees as well. The assessment of the aggregated construct on overall interaction was based on the same research, yet on set of statements. The engagement and predispositions towards the destination image co-creation comprised from three statements (DIC 1–3) on interaction through status (high esteem) and information adapted from the works of Choo and Petrick (2014) and another three on destination loyalty (DIC 4–6) based on Nasir and Yilmaz (2017). Finally, demographic profile scales were developed according to Kani et al. (2017).

Results

Table 1 summarises the socio-demographic characteristics of the studied sample according to the key categorical variables considered: age, income and level of education. On a first reading, the majority of participants were in the age group 36–50, with a medium-income scale (25,001–45,000 Euros) and a higher level of education.

Table 1. Socio-demographic characteristics of the studied sample.

		N	%
Age	18–35	182	35.5
	36–50	227	44.2
	51+	104	20.3
Income (Euros)	25,000	190	37
	25,001–45,000	215	41.9
	45,000+	108	21.1
Education	Primary & Secondary	88	17.2
	Further	139	27.1
	Higher	286	55.8

The descriptive statistics of the 513 (*N*) sample of participants are summarised in Table 2 along the mean, standard deviation, kurtosis and skewness of the sample. Participants seem overall positively inclined towards refugees with the disapproval on acts of discrimination identified as key aspect on their perceptions (RCP 4: 2.77). In terms of marketing influence, tourists who visited Lesvos in the summer of 2019 appear to be drawn mainly by promotional activities undertaken by either tourist agencies or operators (MI 4: 2.33) as well as the perceived brand of the island and its tourism product (MI3: 2.38). Perceptions of safety nor security were leading tourist perceptions on Lesvos destination image prior (PDI 1: 2.79) nor after (CD 1: 2.39) their visit. While on the island tourists shared primarily content on their tourism experience (UGC 4: 1.98) which served directly or indirectly as a type of tourism promotion (UGC 5:1.98). In terms of their interactions, participants identified encounters with other tourists (IT 2: 2.29) and locals (IL 2: 1.87) as primarily friendly. In terms of their encounters with refugees, participants identified their interactions as frequent (IR 4: 2.86) yet harmonious (IR 1:2.83). As for their experience of their overall interactions, participants experienced Lesvos overall as a friendly (OI 1: 2.05) and exciting (OI 4: 2.05) destination, with the latest probably attributed to their whole experience in the island. Finally, the most important constructs of destination image co-creation in terms of locals-tourists interaction was the provision of useful information on tourism services such as transport, attractions, restaurants and accommodation (DIC 2: 2.01) and local culture/lifestyle (DIC 3: 2.2).

Model fit

In order to ensure the goodness of model's fit to the collected dataset, a series of validity indices were calculated. Kline (2010) recommends (a) χ^2, (b) the Comparative Fit Index (CFI), (c) the Root Mean Square Error of Approximation (RMSEA) and, (d) the Standardised Root-Mean-Square Residual (SRMR) as the most appropriate metrics of goodness of fit. Summarising Pappas (2016):

(a) for big sample sizes (in this case N: 513), the goodness of model's fit is assessed as the ratio of the χ^2 statistic against the degrees of freedom (df). Good model fit is supported when $0 \leq \chi^2/\text{df} \leq 2$.
(b) The CFI indicates better fits when is closer to 1.0
(c) The RMSEA indicates close fits for values lower than .5
(d) The SRMR as the square root of the discrepancy between the sample vs. the model covariance matric indicates higher fits for values lower than .8.

A Confirmatory Factor Analysis (CFA) was then employed to identify the most important components of the proposed research model. The CFA results turned a χ^2 value of 711.322 with 378 df ($p < .01$), hence a χ^2/df ratio of 1,882 indicating a good model fit. In terms of the remaining metrics, CFI = .908; RMSEA = .492 and SRMR:.642 ($p < .01$) all suggesting a high model fit. CFA loading coefficients are summarised in Table 3, with absolute values suppressed to the .4 value as recommended by Norman and Streiner (2008). The internal consistency of the model was further assessed through Cronbach's Alpha, giving an overall model reliability value of .789. For individual model constructs, the respective metric values ranged between a min

Table 2. Descriptive statistics.

	Statement	Mean	Std. dev.	Kurtosis	Skewness
Refugee Crisis Perceptions					
RCP 1	Refugees are not infringing destination resources	2.80	0.937	0.332	−1.194
RCP 2	Refugees provide Lesvos with a valuable human resource	2.97	0.796	0.069	−0.769
RCP 3	Refugees should not be forced to go back to their own countries	2.94	0.973	0.147	−1.51
RCP 4	Refugees should not be discriminated against	2.77	0.948	0.524	−1.234
RCP 5	Refugees are not breaking the law	2.82	0.931	0.441	−1.094
Marketing Influence					
MI 1	Direct marketing activities influence my purchasing decisions	3.19	0.857	−0.238	−0.733
MI 2	eWOM influences my purchasing decisions	3.22	0.863	0.057	−0.273
MI 3	The tourism product's brand influences my purchasing decisions	2.38	1.236	0.308	−1.405
MI 4	Promotional activities undertaken by tourist agencies/operators influence my purchasing decisions	2.33	1.145	0.513	−0.983
MI 5	Promotional activities undertaken by destinations influence my purchasing decisions	3.05	0.864	0.326	−0.087
Pre-visit Destination Image					
PDI 1	Before my visit, I perceived Lesvos as a safe destination	2.79	0.613	0.15	−0.513
PDI 2	Before my visit, I perceived Lesvos as a secure destination	2.37	0.541	0.633	−0.288
PDI 3	Before my visit, I perceived Lesvos as a friendly host community	2.12	1.021	0.449	−0.961
PDI 4	Before my visit, I perceived Lesvos to have a helpful host community	2.40	0.741	1.359	0.385
PDI 5	Before my visit, I perceived Lesvos to have a peaceful atmosphere	2.20	0.457	1.554	3.737
Current Destination Image					
CD 1	I perceive that Lesvos is a safe destination	2.39	0.781	0.407	0.557
CD 2	I perceive that Lesvos is a secure destination	2.19	0.637	1.634	4.405
CD 3	I perceive that Lesvos has a friendly host community	1.68	0.647	0.604	0.454
CD 4	I perceive that Lesvos has a helpful host community	1.92	0.528	1.831	11.997
CD 5	I perceive that Lesvos has a peaceful atmosphere	2.11	0.65	1.136	2.391
User-Generated Content					
UGC 1	Whilst being in Lesvos I shared content on the events held on the island	2.55	0.951	0.289	−0.99
UGC 2	Whilst being in Lesvos I shared content on management and protection of stakeholders	2.54	0.928	0.57	−0.966
UGC 3	Whilst being in Lesvos I shared content on tourist safety	2.04	0.893	0.911	0.312
UGC 4	Whilst being in Lesvos I shared content on tourism experience	1.98	0.78	1.056	1.391
UGC 5	Whilst being in Lesvos I shared content on destination promotion	1.98	0.838	1.015	0.868
Interaction with Tourists					
IT 1	My interaction with other tourists was harmonious	2.80	0.626	0.179	−0.582
IT 2	My interaction with other tourists was friendly	2.29	0.5	1.422	1.032
IT 3	My interaction with other tourists was co-operative	2.55	0.623	1.839	8.52
IT 4	My interaction with other tourists was frequent	2.91	1.419	0.886	−0.055
IT 5	My interaction with other tourists was close	3.61	1.416	0.364	−0.954
IT 6	My interaction with other tourists was intense	3.77	1.6	0.416	−1.42
Interaction with Locals					
IL 1	My interaction with locals was harmonious	2.21	0.514	1.556	2.936
IL 2	My interaction with locals was friendly	1.87	0.646	1.263	3.823
IL 3	My interaction with locals was co-operative	2.01	0.826	1.864	7.567
IL 4	My interaction with locals was frequent	2.00	0.715	0.52	0.42
IL 5	My interaction with locals was close	2.03	0.929	1.018	1.632
IL 6	My interaction with locals was intense	2.06	0.993	1.314	2.339
Interaction with Refugees					
IR 1	My interaction with refugees was harmonious	2.83	0.382	−1.852	1.789
IR 2	My interaction with refugees was friendly	3.35	0.487	0.507	−1.479
IR 3	My interaction with refugees was co-operative	4.20	1.366	0.36	−1.517
IR 4	My interaction with refugees was frequent	2.86	0.787	0.261	−1.343
IR 5	My interaction with refugees was close	3.56	0.902	−0.052	0.951
IR 6	My interaction with refugees was intense	2.77	0.692	0.446	−0.028
Overall Interaction					
OI 1	My interaction with other people in Lesvos made me believe that it is a friendly destination	2.05	0.601	0.904	2.555
OI 2	My interaction with other people in Lesvos made me believe that it is a hospitable destination	2.33	0.717	0.094	−0.233

(Continued)

Table 2. Continued.

	Statement	Mean	Std. dev.	Kurtosis	Skewness
OI 3	My interaction with other people in Lesvos made me believe that it is a pleasant destination	2.37	0.612	0.403	0.043
OI 4	My interaction with other people in Lesvos made me believe that it is an exiting destination	2.05	0.601	0.904	2.555
Destination Image Co-creation					
DIC 1	Tourists are treated with high esteem	3.07	1.724	0.957	−0.799
DIC 2	Tourists are provided with useful information, such as transport, attractions, restaurants, and hotels	2.01	0.453	0.911	5.817
DIC 3	Tourists are provided with information on locals' way of life, traditional culture, and local history.	2.20	0.57	1.892	3.77
DIC 4	I will say positive things about Lesvos to other people	2.13	0.627	1.327	2.90
DIC 5	I will encourage friends and relatives to visit Lesvos	2.82	0.704	−1.253	1.826
DIC 6	I will definitely revisit Lesvos	2.76	0.762	−1.159	1.17

.701 for Overall Interaction to max .954 for Refugee Crisis Perceptions, which are all above the min accepted .7 value (Nunnally, 1978). Average Variance Extracted (AVE) for all individual constructs was over .5 suggesting adequate convergent validity levels (Kim et al., 2012), while all constructs Composite Reliability (CR) was again above the recommended .7 acceptance level (Huang et al., 2004).

Hypothesis testing

Figure 2 summarises the hypotheses testing of the model, where all but H4 have been confirmed by the dataset. More specifically, refugee crisis perceptions have a strong positive impact on the pre-visit destination image (H1: $\beta = .274$; $p < .01$). Pre-visit destination image is still positively but only mildly affected by marketing influences (H2: $\beta = .087$; $p < .01$), but significant impact on destination image co-creation (H3: $\beta = .195$; $p < .01$). Yet, the latest is strongly affected by the actual destination image tourist formulate during the visit at the destination (H5: $\beta = .315$; $p < .01$) and even more importantly by the UGC they produce during their stay (H6: $\beta = .398$; $p < .01$). The construct of overall interaction is affected by all three identified group encounters. Tourist to tourist interactions have a mild yet positive contribution (H7: $\beta = .092$; $p < .01$) to overall interactions, as well as interactions with locals which score higher and still positively (H8: $\beta = .120$; $p < .01$). Interactions with the third ingroup of refugees score the highest in terms of significance amongst the other groups, yet with a negative contribution to the overall image (H9: $\beta = -.287$; $p < .01$). Finally, the perception of overall interactions on destination has the strongest significance towards the destination image co-creation (H10: $\beta = .402$; $p < .01$). Interestingly enough, the only hypothesis not confirmed by the data on Lesvos was the one suggesting that pre-visit destination image has a significant effect on the actual destination image tourists formulate during their visit (H4: $\beta = .227$; $p > .05$).

Discussion

The research focused on the perspectives and attitudes of tourists who actually decided to proceed with their visit (not mere intentions) and their contribution towards the co-

Table 3. Validity and reliability analysis.

	Loadings	*A*	AVE	CR
Refugee Crisis Perceptions		0.954	0.684	0.896
RCP 1	0.829			
RCP2				
RCP 3	0.828			
RCP 4	0.868			
RCP 5	0.781			
Marketing Influence		0.845	0.583	0.873
MI 1	0.891			
MI 2	0.822			
MI 3	0.681			
MI 4	0.648			
MI 5	0.748			
Pre-visit Destination Image		0.739	0.560	0.864
PDI 1	0.791			
PDI 2	0.722			
PDI 3	0.681			
PDI 4	0.746			
PDI 5	0.794			
Current Destination Image		0.743	0.690	0.899
CD 1	0.798			
CD 2	0.829			
CD 3	0.874			
CD 4				
CD 5	0.821			
User-Generated Content		0.852	0.571	0.866
UGC 1	0.503			
UGC 2	0.832			
UGC 3	0.709			
UGC 4	0.822			
UGC 5	0.855			
Interaction with Tourists		0.739	0.837	0.953
IT 1				
IT 2				
IT 3	0.785			
IT 4	0.903			
IT 5	0.982			
IT 6	0.976			
Interaction with Locals		0.720	0.568	0.840
IL 1				
IL 2				
IL 3	0.778			
IL 4	0.769			
IL 5	0.751			
IL 6	0.715			
Interaction with Refugees		0.756	0.541	0.823
IR 1				
IR 2	0.608			
IR 3	0.806			
IR 4	0.706			
IR 5	0.804			
IR 6				
Overall Interaction				
OI 1	0.833	0.701	0.623	C.868
OI 2	0.724			
OI 3	0.785			
OI 4	0.812			
Destination Image Co-creation		0.767	0.577	0.871
DIC 1	0.718			
DIC 2	0.832			
DIC 3	0.815			
DIC 4				
DIC 5	0.718			
DIC 6	0.707			

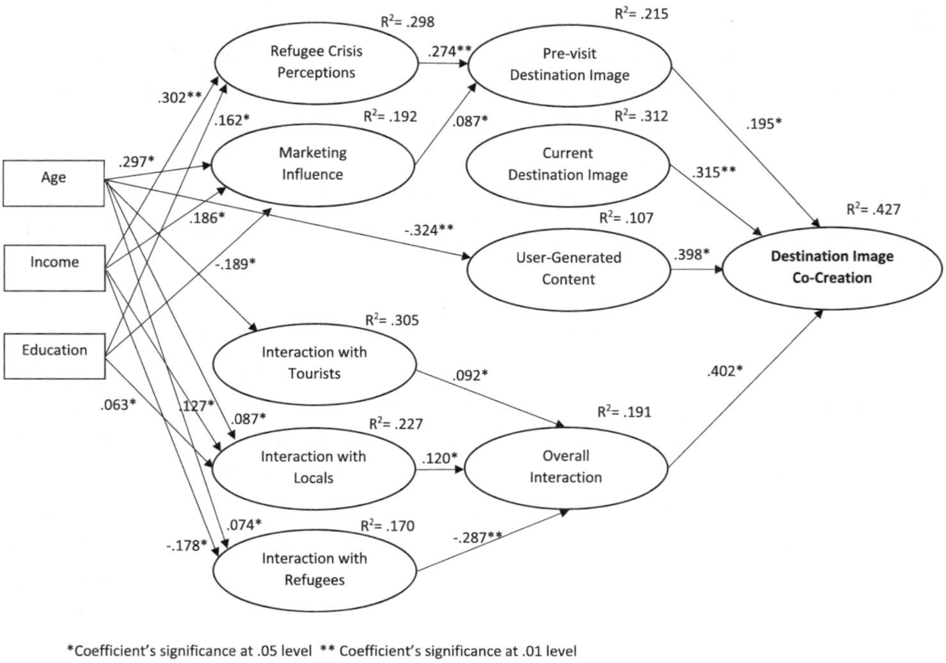

Figure 2. Destination image co-creation model. *Coefficient's significance at .05 level, **Coefficient's significance at .01 level.

creation of a destinations' image, regardless or in addition to all other efforts taken by other stakeholder parties (locals, DMOs, TOs etc.). Research findings suggest the importance of convincing tourists to proceed with their purchasing (visiting) intention as, expectedly, their pre-visit destination image has the lowest significance to their image co-creation formulation. Okuyama's (2018) analysis on the timing of demand recovery policies seems to resonate even for destinations in prolonged crisis.

Amongst the pre-visit perceptions' exploration statements, the affective image of Lesvos as a friendly and helpful host community (PDI 3-PDI 4) were the ones who possibly triggered visitation regardless of tourist's safety and security concerns. In regard to the constructs actually informing the pre-visit image of a destination affected by ingroup mobilities, findings suggest a strong positive attitude bias from tourists' perception on the refugee crisis overall. The latest confirm Legg et al. (2012) and Simpson et al. (2016) suggestions that ideological beliefs other that influencing ones' viewing of the world have very strong implications for their purchasing behaviours particularly during times of political and social crises. Marketing influences had a much lower contribution to the pre-visit destination image, with the actual strength of the brand (MI 3) and promotional activities from TOs (MI 4) leading on their intentions. This is somehow expected under "normal" destination circumstances (Morgan & Pritchard, 2004), yet it was interesting to observe in destinations in sustained crisis. The latest probably suggests that tourism behavioural tendencies bounce back after a certain period of familiarisation with a crisis and minimisation of the elements of relates surprise. Findings confirm Pappas (2019) suggestion of the trust on the brand value yet differ on the importance of eWOM towards purchasing intentions. The reliability and credibility of the information sources does not seem to play an important role in times of sustained crisis as it could be considered biased, hence official promotional activities from destinations didn't seem to contribute much on the formulation of destination image prior to the visit. It is interesting to see that direct marketing activities (MI 1) and eWOM (MI 2) scored the lowest amongst the constructs explored, due to probably being considered as biased and unreliable (Kotler et al., 2014).

While at the destination, their cognitive image and possibly overall experience seems to have enhanced their appreciation and empathy for the local host community (CD 3–CD 4) and reassured their concerns of threat and turbulence, resulting to much higher (in comparison to pre-visit) significance of contribution towards image co-creation. These findings complement previous research from Tasci and Gartner (2007) on the functionality of relationships in destinations and Tucker's (2016) on tourism empathy. Even more significant for the destination image co-creation turned to be tourists' contribution and engagement with UGC during their visit, primarily anything related to tourist's experiences in the island (UGC 4) promoting consciously or unconsciously the destination (UGC 5) and without necessarily referring to any actual events taking place or the implications for the involved stakeholders (UGC 1–2). These findings differ from previous research on the communication and generation of information intelligence during disasters (Sigala, 2011; Takahashi et al., 2015) yet, align more with the post-crisis discourse (Okuyama, 2018; Ulmer et al., 2007). It seems that as long as there is no obvious or imminent crisis tourists perceived as rather past. Such attituded trigger tourist destination image co-creation via sharing positive feedback of their experience on the destination, still they don't seem strong enough to convince them necessarily to repeat the visit (DIC 6) nor directly recommend the destination (DIC 5).

Overall interaction was the most significant construct directly related to destination image co-creation, indicating the importance of interpersonal encounters to generate empathy and affective attachment to the visited destination (Tucker, 2016; Yang, 2016). In exploring the significance of the categorical demographic attributes of age, income and education, findings of this study suggest age as the highest underlying contribution for engagement in the destination image co-creation, followed by income and later education. These findings are overall in line with smart technologies usage trends (Wang et al., 2016).

Conclusions

This research builds on a sample of domestic and international tourists who visited the island of Lesvos (Greece) seven years after the beginning of the refugee and immigrant mobility in the Mediterranean region. The study aimed to explore their cognitive and affective predispositions towards the affected destination and to identify attitudinal patterns and triggers for engagement in destination image co-creation. Findings suggest that any perceptual fears prior to the visit, which are usually triggered by one's personal ideological beliefs on ingroup mobilities overall, are relaxed once they decide to visit the destination. It is then both the cognitive image of the actual situation as well that the affective one generated through their encounters at destinations that leads their empathetic behaviours towards both the ingroup and outgroup stakeholders and engage them into sharing their perceived value of their overall experience. At this point of the crisis cycle, tourists preferred to share on their actual experience and attributes of the destination rather than on actual crisis-related events or opinions on involved stakeholders. The latest is optimum for the indirect marketing of affected destinations, that continue to demonstrate their product and service experience through organic and experiential channels. Regardless of the duration of the crisis, research findings confirm the importance of tourists' engagement in a destination image co-creation which remains subject to safeguarding and ensuring the actual visitation intension.

The concept of co-creation has been applied along the service-dominant approach which allows for the exploration of the multiple attitudes and experience facets that trigger tourists' response and engagement in the co-creation of a destination's image. In that regard, it contributes to the advancement of its conceptualisation through the exploration of additional control variables (UGC, overall interactions). In view of the research body on crisis cycles and their implications for a destination's recovery though, the research contributes to the exploration of the intensity and duration effect and triggers of the image of destinations in sustained crisis. The managerial implications of this research hence relate to the response and recovery phase of destinations' crisis management, and primarily to the identification of the optimum timing and channels of interventions, which only acquires particular importance in the context of the ongoing population mobilities at global level.

Findings reflect the particularities of Lesvos as a destination and should be analysed bearing in mind both the actual refugee/immigrant management initiatives in the island over the last years (location; pressure; carrying capacity) and the characteristics of tourist product and island clientele. Lesvos is not a tourism destination in the traditional term of the context, nor has a distinctive tourism destination brand. Findings of the research are useful to enhance the underlying dynamics of tourist behaviours and attitudes at similar destination, yet they should be cautiously generalised within the specifics of the tourist population of the island.

Future research should consider more in detail the implications of client profile on the image co-creation process, by including attributes of e.g. familiarity to the destination (second housing, VFR, second generation Greeks living overseas); nationality (media coverage); tourism product; or ideological orientation. The current research didn't consider necessary to fragment the analysis, as the research aim revolved around the identification of external (destination-induced) rather than internal (psychographic and behavioural) triggers to response and engagement. Obviously, such a delineation would offer another layer of perceptual analysis and adaptation fit to other destinations. The paper here adds a small stone in this larger endeavour. Moreover, the discussion over the ingroup's management is beyond the scope of this paper, yet any future relevant research should better differentiate between perceptions over war refugees and undocumented immigrants.

Disclosure statement

No potential conflict of interest was reported by the author(s).

Data availability statement

Data set available upon request.

References

Ajzen, I. (1991). The theory of planned behavior. *Organizational Behavior and Human Decision Processes*, *50*(2), 179–211. http://doi.org/10.1016/0749-5978

Atwell Seate, A., & Mastro, D. (2015). Exposure to immigration in the news: The impact of group-level emotions on intergroup behavior. *Communication Research*, *44*(6).

Baloglu, S., & McCleary, K. W. (1999). A model of destination image formation. *Annals of Tourism Research*, *26*(4), 868–897. https://doi.org/10.1016/S0160-7383(99)00030-4

Bauer, R. A. (1960). Consumer behaviour as risk taking. In R. Hancock (Ed.), *Dynamic marketing for a changing world* (pp. 389–398). American Marketing Association.

Bierbrauer, G., & Klinger, E. W. (2002). Political ideology, perceived threat and justice toward immigrants. *Social Justice Research*, *15*(1), 41–52. https://doi.org/10.1023/A:1016045731732

Blumer, H. (1958). Race prejudice as a sense of group position. *Pacific Sociological Review*, *1*(1), 3–7. https://doi.org/10.2307/1388607

Chen, C.-F., & Chen, F.-S. (2010). Experience quality, perceived value, satisfaction and behavioural intentions for heritage tourists. *Tourism Management*, *31*, 9–35. https://doi.org/10.1016/j.tourman.2009.02.008.

Chen, C. F., & Tsai, D. (2007). How destination image and evaluative factors affect behavioural intentions? *Tourism Management*, *28*(4), 1115–1122. https://doi.org/10.1016/j.tourman.2006.07.007

Chew, E. Y. T., & Jahari, S. A. (2014). Destination image as a mediator between perceived risks and revisit intention: A case of post-disaster Japan. *Tourism Management*, *40*, 382–393. http://doi.org/10.1016/j.tourman.2013.07.008

Choo, H., & Petrick, J. F. (2014). Social interactions and intentions to revisit for agritourism service encounters. *Tourism Management*, *40*, 372–381. https://doi.org/10.1016/j.tourman.2013.07.011

Cirer-Costa, J. C. (2017). Turbulence in Mediterranean tourism. *Tourism Management Perspectives*, *22*, 27–33. http://doi.org/10.1016/j.tmp.2017.01.004

Coombs, W. T. (2015). The value of communication during a crisis: Insights from strategic communication research. *Business Horizons*, *58*(2), 141–148. https://doi.org/10.1016/j.bushor.2014.10.003

Croucher, S. M. (2013). Integrated threat theory and acceptance of immigrant assimilation: An analysis of Muslim immigration in Western Europe. *Communication Monographs*, *80*(1), 46–62. https://doi.org/10.1080/03637751.2012.739704

Deutsche Welle. (2020). Τι θα γίνει τώρα με τους νέους πρόσφυγες στη Λέσβο. Greek. Retrieved April 17, 2020, from https://www.dw.com/el/τι-θα-γίνει-τώρα-με-τους-νέους-πρόσφυγες-στη-λέσβο/a-52636235

Duverger, P. (2013). Curvilinear effects of user-generated content on hotels' market share. *Journal of Travel Research*, *52*(4), 465–478. http://doi.org/10.1177/0047287513478498

Farmaki, A., & Christou, P. (2019). Refugee migration and service industries: Advancing the research agenda. *The Service Industries Journal*, *39*(9–10), 668–683. https://doi.org/10.1080/02642069.2018.1435643

Filieri, R. (2015). What makes online reviews helpful? A diagnosticity-adoption framework to explain informational and normative influences in e-WOM. *Journal of Business Research*, *68*(6), 1261–1270. http://doi.org/10.1016/j.jbusres.2014.11.006

Fuchs, G., Uriely, N., Reichel, A., & Maoz, D. (2013). Vacationing in a terror-stricken destination: Tourists' risk perceptions and rationalizations. *Journal of Travel Research*, *52*(2), 182–191. https://doi.org/10.1177/0047287512458833

Gartner, W. C. (1994). Image formation process. *Journal of Travel and Tourism Marketing*, *2*(2/3), 191–215. https://doi.org/10.1300/J073v02n02_12

George, R. (2010). Visitor perceptions of crime-safety and attitudes towards risk: The case of Table Mountain National Park, Cape Town. *Tourism Management*, *31*(6), 806–815. https://doi.org/10.1016/j.tourman.2009.08.011

Gross, M. J., & Brown, G. (2008). An empirical structural model of tourists and places: Progressing involvement and place attachment into tourism. *Tourism Management*, *29*(6), 1141–1151. http://doi.org/10.1016/j.tourman.2008.02.009

Huang, J., & Hsu, C. H. C. (2010). The impact of customer-to-customer interaction on cruise experience and vacation satisfaction. *Journal of Travel Research*, *49*(1), 79–92. http://doi.org/10.1177/0047287509336466

Huang, W., Schrank, H., & Dubinsky, A. J. (2004). Effect of brand name on consumers' risk perceptions of online shopping. *Journal of Consumer Behaviour*, *4*(1), 40–50. http://doi.org/10.1002/

Hudson, S., & Thal, K. (2013). The impact of social media on the consumer decision process: Implications for tourism marketing. *Journal of Travel & Tourism Marketing*, *30*(1-2), 156–160. http://doi.org/10.1080/10548408.2013.751276

Kani, Y., Aziz, Y. A., Sambasivan, M., & Bojei, J. (2017). Antecedents and outcomes of destination image of Malaysia. *Journal of Hospitality and Tourism Management*, *32*, 89–98. https://doi.org/10.1016/j.jhtm.2017.05.001

Kapferer, J. N. (2008). *New strategic brand management: Creating and sustaining brand equity long term*. Kogan Page.

Kaplan, L. B., Szybillo, G. J., & Jacoby, J. (1974). Components of perceived risk in product purchase: A cross-validation. *Journal of Applied Psychology*, *59*(3), 287–291. https://doi.org/10.1037/h0036657

Khan, M. J., Chelliah, S., & Ahmed, S. (2017). Intention to visit India among potential travellers: Role of travel motivation, perceived travel risks, and travel constraints. *Tourism and Hospitality Research*, *19*(3), 351–367. http://doi.org/10.1177/1467358417751025

Kim, H. W., Xu, Y., & Gupta, S. (2012). Which is more important in Internet shopping, perceived price or trust? *Electronic Commerce Research and Applications*, *11*(3), 241–252. http://doi.org/10.1016/j.elerap.2011.06.003

Kline, R. B. (2010). *Principles and practice for structural equation modelling*. Guildford Press.

Kock, F., Joiassen, A., & Assaf, A. G. (2016). Advancing destination image: The destination content model. *Annals of Tourism Research*, *61*, 28–44. https://doi.org/10.1016/j.annals.2016.07.003

Kotler, P. T., Bowen, J. T., Makens, J., & Baloglu, S. (2017). *Marketing for hospitality and tourism* (7th ed.). Pearson Education.

Kotler, P., Haider, D. H., & Rein, I. (1993). *Marketing places: Attracting investment, industry, and tourism to cities, states and nations*. Free Press.

Kotler, P., Keller, K. L., Ancarani, F., & Costabile, M. (2014). *Marketing management* (14th ed.). Pearson.

Kozak, M., Crotts, J. C., & Law, R. (2007). The impact of the perception of risk on international travellers. *International Journal of Tourism Research*, *9*(4), 233–242. https://doi.org/10.1002/jtr.607

Lamsfus, C., Wang, D., Alzua-Sorzabal, A., & Xiang, Z. (2014). Going mobile defining context for on-the-go travelers. *Journal of Travel Research*, *54*(6), 691–701. http://doi.org/10.1177/0047287514538839

Lee, R. M. (1993). *Doing research on sensitive topics*. Sage.

Lee, Z., & Sargeant, A. (2011). Dealing with social desirability bias: An application to charitable giving. *European Journal of Marketing*, *45*(5), 703–719. https://doi.org/10.1108/03090561111119994

Legg, M. P., Tang, C. H. H., & Slevitch, L. (2012). Does political ideology play a role in destination choice? *American Journal of Tourism Research*, *1*(2), 45–58. http://worldscholars.org/index.php/ajtr/article/view/148.

Loi, K. I., & Pearce, P. L. (2012). Annoying tourist behaviours: Perspectives of hosts and tourists in Macao. *Journal of China Tourism Research*, *8*(4), 395–416. https://doi.org/10.1080/19388160.2012.729411

Lusch, R. F., & Vargo, S. L. (2006). Service-dominant logic: Reactions, reflections and refinements. *Marketing Theory*, *6*(3), 281–288. https://doi.org/10.1177/1470593106066781

Mair, J., Ritchie, B. W., & Walters, G. (2016). Towards a research agenda for post-disaster and post-crisis recovery strategies for tourist destinations: A narrative review. *Current Issues in Tourism*, *19*(1), 1–26. https://doi.org/10.1080/13683500.2014.932758

Matzler, K., Strobl, A., Stockburger-Sauer, N., Bobovnicky, A., & Bauer, F. (2016). Brand personality and culture: The role of cultural differences on the impact of brand personality perceptions on tourists' visit intentions. *Tourism Management*, *52*, 507–520. https://doi.org/10.1016/j.tourman.2015.07.017

Morgan, N., & Pritchard, A. (2004). Meeting the destination branding challenge. In N. Morgan, A. Pritchard, & R. Pride (Eds.), *Destination branding: Creating the unique destination proposition* (pp. 59–79). Elsevier.

Nasir, S., & Yilmaz, M. T. (2017). Vacationing at a destination under terrorism risk: Tourists' destination image perceptions about Istanbul. *The Online Journal of Science and Technology*, *7*(1), 139–145.

Norman, G., & Streiner, D. (2008). *Biostatistics: The bare essentials* (3rd ed.). Hamilton.

Nunnally, J. C. (1978). *Psychometric theory*. McGraw-Hill.

Okuyama, T. (2018). Analysis of optimal timing of tourism demand recovery policies from natural disaster using the contingent behavior method. *Tourism Management*, *64*, 37–54. https://doi.org/10.1016/j.tourman.2017.07.019

Oliveira, A., & Huertas, A. (2019). How do destination use Twitter to recover their images after a terrorist attack. *Journal of Destination Marketing and Management*, *12*, 46–54. https://doi.org/10.1016/j.jdmm.2019.03.002

Ortiz, C. M. S., Frias-Jamilena, D. M., & Garcia, J. A. C. (2015). Overall perceived value of a tourism service: Analysing the spillover effect between electronic channel and consumption of the hotel service. *Tourism and Hospitality Research*, 1–11. https://doi.org/10.1177/1467358415613410.

Pappas, N. (2016). Marketing strategies, perceived risks, and consumer trust in online buying behavior. *Journal of Retailing and Consumer Services*, *29*, 92–103. https://doi.org/10.1016/j.jretconser.2015.11.007

Pappas, N. (2019). The complexity of consumer experience formulation in the sharing economy. *International Journal of Hospitality Management*, *77*, 415–424. https://doi.org/10.1016/j.ijhm.2018.08.005

Pappas, N., & Papatheodorou, A. (2017). Tourism and the refugee crisis in Greece: Perceptions and decision-making of accommodation providers. *Tourism Management*, *63*, 31–41. https://doi.org/10.1016/j.tourman.2017.06.005

Paraskevas, A., & Arendell, B. (2007). A strategic framework for terrorism prevention and mitigation in tourism destinations. *Tourism Management*, *28*(6), 1560–1573. https://doi.org/10.1016/j.tourman.2007.02.012

Pearce, P. L. (2005). *Tourist behaviour: Themes and conceptual schemes*. Channel View Publications.

Pine, J., & Gilmore, J. (1998, July–August). Welcome to the experience economy. *Harvard Business Review*, 97–105.

Plog, S. C. (2016). Why destination areas rise and fall in popularity. *Cornell Hotel and Restaurant Administration Quarterly*, *14*(4), 55–58. http://doi.org/10.1177/001088047401400409

Qu, H., Kim, L. H., & Im, H. H. (2011). A model of destination branding: Integrating the concepts of the branding and destination image. *Tourism Management*, *32*(3), 465–476. http://doi.org/10.1016/j.tourman.2010.03.014

Quintal, V. A., Lee, J. A., & Soutar, G. N. (2010). Risk, uncertainty and the theory of planned behavior: A tourism example. *Tourism Management*, *31*(6), 797–805. http://doi.org/10.1016/j.tourman.2009.08.006

Ramseook-Munhurrun, P., Seebaluck, V. N., & Naidoo, P. (2015). Examining the structural relationships of destination image, perceived value, tourist satisfaction and loyalty: Case of Mauritius. *Procedia – Social and Behavioral Sciences*, *175*, 252–259. https://doi.org/10.1016/j.sbspro.2015.01.1198

Rodríguez Molina, M. Á., Frías-Jamilena, D.-M., & Castañeda-García, J. A. (2013). The moderating role of past experience in the formation of a tourist destination's image and in tourists' behavioural intentions. *Current Issues in Tourism*, *16*(2), 107–127. http://doi.org/10.1080/13683500.2012.665045

Seetaram, N. (2012). Immigration and international inbound tourism: Empirical evidence from Australia. *Tourism Management*, *33*(6), 1535–1543. https://doi.org/10.1016/j.tourman.2012.02.010

Sigala, M. (2011). Social media and crisis management in tourism: Applications and implications for research. *Information Technology & Tourism*, *13*(4), 269–283. https://doi.org/10.3727/109830512X13364362859812

Simpson, J. J., Simpson, P. M., & Cruz-Milan, O. (2016). Attitude towards immigrants and security: Effects on destination-loyal tourists. *Tourism Management*, *57*, 373–386. https://doi.org/10.1016/j.tourman.2016.06.021

Sönmez, S. F., Apostolopoulos, Y., & Tarlow, P. (2016). Tourism in crisis: Managing the effects of terrorism. *Journal of Travel Research*, *38*(1), 13–18. http://doi.org/10.1177/004728759903800104

Sönmez, S. F., & Graefe, A. R. (1998). Determining future travel behaviour from past travel experience and perceptions of risk and safety. *Journal of Travel Research*, *37*(2), 171–177. https://doi.org/10.1177/004728759803700209.

Stephan, W. G. (2014). Intergroup anxiety: Theory, research, and practice. *Personality and Social Psychology Review*, *18*(3), 239–255. https://doi.org/10.1177/1088868314530518

Stylidis, D., Shani, A., & Belhassen, Y. (2017). Testing an integrated destination image model across residents and tourists. *Tourism Management*, *58*, 184–195. http://doi.org/10.1016/j.tourman.2016.10.014

Stylos, N., Bellou, V., Andronikidis, A., & Vassiliadis, C. A. (2017). Linking the dots among destination images, place attachment, and revisit intentions: A study among British and Russian tourists. *Tourism Management*, *60*, 15–29. https://doi.org/10.1016/j.tourman.2016.11.006

Suntikul, W., & Jachna, T. (2016). The co-creation/place attachment nexus. *Tourism Management*, *52*, 276–286. https://doi.org/10.1016/j.tourman.2015.06.026

Takahashi, B., Tandoc, E. C., & Carmichael, C. (2015). Communicating on twitter during a disaster: An analysis of tweets during typhoon Haiyan in the Philippines. *Computers in Human Behavior*, *50*, 392–398. https://doi.org/10.1016/j.chb.2015.04.020

Tan, W.-K., & Wu, C.-E. (2016). An investigation of the relationships among destination familiarity, destination image and future visit intention. *Journal of Destination Marketing & Management*, *5*(3), 214–226. http://doi.org/10.1016/j.jdmm.2015.12.008

Tasci, A. D., & Gartner, W. C. (2007). Destination image and its functional relationships. *Journal of Travel Research*, *45*(4), 413–425. https://doi.org/10.1177/0047287507299569

Tsartas, P., Kyriakaki, A., Stavrinoudis, T., Despotaki, G., Doumi, M., Sarantakou, E., & Tsilimpokos, K. (2019). Refugees and tourism: A case study from the islands of Chios and Lesvos, Greece. *Current Issues in Tourism*, *23*(11), 1311–1327. https://doi.org/10.1080/13683500.2019.1632275.

Tucker, H. (2016). Empathy and tourism: Limits and possibilities. *Annals of Tourism Research*, *57*, 31–43. https://doi.org/10.1016/j.annals.2015.12.001

Tussyadiah, I. P. (2015). The influence of innovativeness on on-site smartphone use among American travelers: Implications for context-based push marketing. *Journal of Travel & Tourism Marketing*, *33*(6), 806–823. http://doi.org/10.1080/10548408.2015.1068263

Ulmer, R., Seeger, M., & Sellnow, T. (2007). Post-crisis communication and renewal: Expanding the parameters of post-crisis discourse. *Public Relations Review*, *33*(2), 130–134. https://doi.org/10.1016/j.pubrev.2006.11.015

Uşaklı, A., Koç, B., & Sönmez, S. (2017). How 'social' are destinations? Examining European DMO social media usage. *Journal of Destination Marketing & Management*, *6*(2), 136–149. http://doi.org/10.1016/j.jdmm.2017.02.001

Waligo, V. (2013). Great expectations: Imagination and anticipation in tourism. *Current Issues in Tourism*, *16*(5), 514–515. https://doi.org/10.1080/13683500.2012.754846

Wang, D., Xiang, Z., & Fesenmaier, D. R. (2016). Smartphone use in everyday life and travel. *Journal of Travel Research*, *55*(1), 52–63. https://doi.org/10.1177/0047287514535847

White, C. J. (2014). Ideal standards and attitude formation: A tourism destination perspective. *International Journal of Tourism Research*, *16*(5), 441–449. http://doi.org/10.1002/jtr.1938

Williams, N. W., Inversini, A., Ferdinand, N., & Buhalis, D. (2017). Destination eWOM: A macro and meso network approach? *Annals of Tourism Research*, *64*, 87–101. https://doi.org/10.1016/j.annals.2017.02.007

Woosnam, K. M., Shafer, C. S., Scott, D., & Timothy, D. J. (2015). Tourists' perceived safety through emotional solidarity with residents in two Mexico–United States border regions. *Tourism Management*, *46*, 263–273. http://doi.org/10.1016/j.tourman.2014.06.022

Xu, F., La, L., Zhen, F., Lobsang, T., & Huang, C. (2019). A data-driven approach to guest experiences and satisfaction in sharing. *Journal of Travel & Tourism Marketing*, *36*(4), 484–496. https://doi.org/10.1080/10548408.2019.1570420

Yang, F. X. (2016). Tourist co-created destination image. *Journal of Travel and Tourism Marketing*, *33*(4), 425–439. https://doi.org/10.1080/10548408.2015.1064063

Ye, S., Soutar, G. N., Sneddon, J. N., & Lee, J. A. (2017). Personal values and the theory of planned behavior: A study of values and holiday trade-offs in young adults. *Tourism Management*, *62*, 107–109. https://doi.org/10.1016/j.tourman.2016.12.023

Yüksel, A., & Yüksel, F. (2007). Shopping risk perceptions: Effects on tourists' emotions, satisfaction and expressed loyalty intentions. *Tourism Management*, *28*(3), 703–713. https://doi.org/10.1016/j.tourman.2006.04.025

Zenker, S., von Wallpach, S., Braun, E., & Vallaster, C. (2019). How the refugee crisis impacts the decision structure of tourists: A cross-country scenario study. *Tourism Management*, *71*, 197–212. https://doi.org/10.1016/j.tourman.2018.10.015

Zhang, H., Fu, X., Cai, L. A., & Lu, L. (2014). Destination image and tourist loyalty: A meta-analysis. *Tourism Management*, *40*, 213–223. https://doi.org/10.1016/j.tourman.2013.06.006

Co-creating Value in Desert Tourism Experiences

Eleni Michopoulou, Idrees Al-Qasmi and Claudia Melpignano

ABSTRACT
This study investigates the determinants of value co-creation in desert camps in Oman from both the customers' and the camp managers' perspectives. The concept of value co-creation in hospitality and tourism has been investigated in a range of ways in the extant literature. However, limited attention has been paid in the process of value co-creation in remote and unique destinations such as desert camps. This research focuses on 5 aspects of value co-creation which are then explored both quantitatively and qualitatively. The findings of the study indicate that within the context of desert camps, value co-creation is influenced by authenticity, engagement, place attachment, and marketing though the value-in-use concept. However, the level of this influence varies between the customers and the camp managers. Finally, findings are discussed in the light of this variance to identify and provide recommendations that enhance value co-creation in the desert camps of Oman.

Introduction

Recognised in recent years as the key to success in a competitive market environment (Gouillart & Ramaswamy, 2014; Pappas & Michopoulou, 2019), the value co-creation concept has been applied in many areas of the tourism and hospitality industry (Buhalis & Foerste, 2015). The importance of embracing the value co-creation concept in the hospitality and service industry arises from the fact that customers view service as means to facilitate the creation of their own experiences (Agrawal & Rahman, 2013). In this context, Bimonte and Punzo (2016) argue that the development of tourism services and products requires adopting a strategy where the different parties are involved in the process of co-creating value (Prebensen et al., 2016). In fact, value co-creation is based on the notion that business organisations are no longer the only arbiters of value and that customers should not be passive actors (Prahalad & Ramaswamy, 2004; Prebensen & Xie, 2017), but operant resources in the value co-creation process (Vargo & Lusch, 2014). As such, value co-creation helps service providers to better understand their customers' increasing sophistication through allowing them to shape aspects of the products and services (Ketonen-Oksi et al., 2016). While there is considerable knowledge of the value co-creation concept in general, little is known about how this concept can be practised within the context of unique destination configurations such as desert camps (Mesoudi et al., 2018).

Context of the study

The tourism sector in Oman has recently witnessed significant growth (MTO, 2018). According to the report of the World Economic Forum, Oman is among the top ten fastest growing destinations in the world for leisure-travel spending between 2016 and 2026 (WEF, 2018). The growth of tourism sector in Oman is estimated to reach 11 million tourists by 2040 (FSG, 2018). The direct added value of the tourism sector in Oman rose to 2.3 billion USD in 2019 (TOF, 2020). Oman's history, culture and heritage constitute some of the major tourist attractions in Oman (Henderson, 2015), but its environment and, more specifically its desert, constitute the main tourism resources (MTO, 2018). Desert tourism and desert's camps provide a solid base to many economical activities. In addition, desert's camps contribute to generate many jobs like local guides, local professional sand-dune drivers and local handcrafts artists. According to the Ministry of Tourism of Oman, 117,000 guests have spent at least one overnight in Eastern AL-Shargiyah Region which hosts most of the desert camps in Oman during the touristic season of 2018 (MTO, 2018).

The desert camp is a type of accommodation facility that offers a wide range of services and activities for desert guests (Cooke, 2010). There are 20 desert camps in Oman which have different capacities of rooms that range from 20 up to 67 room per camp (MTO, 2020). These camps vary from basic camps offering meals and shelter to luxury five-star camps that offer high-end services and facilities (Mesoudi et al., 2018). These camps target mainly international tourists with cultural and natural interests. The camps are located in natural area where several activities can be performed which include Dune Bashing by 4WD vehicles, dinner under the stars, sandboarding, quad biking, camels riding, Bedouin folkloric music, serving typical dishes from the Omani Cuisine, sunrise and sunset watching in the desert, guided desert tours and discovering the local community by visiting Bedouin houses (MTO, 2018).

Literature on desert tourism is scant, and studies adhere mostly to sustainable tourism and eco-tourism discourses (Baker & Mearns, 2017; Jangra & Kaushik, 2018; Pérez-Liu & Tejada-Tejada, 2020; Tremblay, 2006; Woyo & Amadhila, 2018). However, there has been some research focusing on the tourism aspect of the desert experience, and although limited, it addresses different features of the experience. Some notable examples include the works by Narayanan and Macbeth (2009, p. 371) who focused on four-wheel drive experiences in the Australian deserts, noting "the euphoria and meaningfulness in travelling through the seemingly unending open spaces". Moufakkir and Selmi (2018) also looked at the spirituality of the Sahara desert tourism experience, and Allan (2016) focused of the concept of place attachment at the Wadi Rum desert in Jordan. Chatty (2016), in particular, discussed the role of perceived authenticity of desert tourism in Oman, examining how the concept is contested by different social actors.

However, desert tourism (Cooke, 2010) and its development strategies in Oman (Mesoudi et al. 2018) have received relatively little attention in the literature compared to other forms of tourism. Based on this, the study contributes to the growing need to investigate various aspects of the tourism sector in Oman through analysing the concept of value co-creation in the desert camps. To do so, this study explored and reviewed several models of value co-creation (Galvagno & Dalli, 2014; Grönroos, 2011;

Prahalad & Ramaswamy, 2004), and identified the most appropriate to subsequently provide solutions for the critical issues which limit value co-creation in the Omani desert camps. The study uncovers the role of customers and the desert camp managers in the process of value co-creation and provides camp managers with guidelines to enhance their services through identifying the dominant factors in value co-creation from the customer's perspective.

Theoretical background

Value and value co-Creation in tourism

The literature shows that the concepts of creating products and services with values for customers have recently witnessed radical changes (Prebensen & Xie, 2017). In this context, Gilmore and Pine (2015) argue that the process of production has been shifted towards the customisation process, where customers become more engaged in the process of creating and enhancing their experiences. According to Prahalad and Ramaswamy (2004)'s view, companies can gain advantages in a competitive market environment, if they move towards a value-centric approach that allows for the co-creation of values with customers.

Within the service industry specifically, value for customers is a positive feeling that can be obtained when using a certain product or service (Grönroos, 2011; Agrawal & Rahman, 2013). Grönroos (2011) also forwards the perception of service as a concept based on creating value, rather than a type of market offering. Value creation is therefore come to be considered as the core of the service industry in the current marketplace (Boksberger & Melsen, 2011).

Value co-creation is derived from the service-dominant logic concept which is based on an exchange of resources and services between parties to create a shared value (Vargo & Lusch, 2014), based on the social, cultural, economic, an environmental mechanisms of all actors (Cannas, 2018). Revolving around the idea that it contributes to bring multiple stakeholders together onto a single platform (Lugosi, 2014), value co-creation is also defined as an organisation's ability to embrace active contributions from consumers during the process of creating value-added products and services (Hsiao et al., 2015).

The emergence of value co-creation was influenced by changing aspects of the marketplace, in particular by technology and by the customers' changing expectations (Chathoth et al., 2016). Arguably, the incorporation of customers' needs and expectations across the value chain and the subsequent emergence of the value co-creation concept (Im & Qu, 2017) was facilitated through IT-innovation (Franke & Schreier, 2010). Also, the increasing desire of a guest to capture a value from using services in the hospitality industry has encouraged them to take an active role in designing their experience (Munar & Jacobsen, 2014).

Also, the concept of loyalty has changed from customers being loyal to certain organisations to customers becoming loyal to values (Gilmore & Pine, 2015). In fact, studies prove that the guests' loyalty and satisfaction in the tourism and hospitality industry are linked to their ability to influence the value creation process so that it meets their needs, wants and expectations (Luo et al., 2015; Mathis et al., 2016). Businesses operating in the tourism industry should therefore view value co-creation as a mechanism which can help gaining advantage in the competition process (Chesbrough, 2010), as it identifies and specifies the target market's segments and constructs a reliable structure for the value chain (Hsiao et al., 2015). As value co-creation attempts to create experiences for customers that truthfully fulfil their needs and wants (Gilmore & Pine, 2015), it is argued that organisations should design their attractions in ways that enable customers to co-create their experiences (Grayson & Martinec, 2004).

Aspects of value co-creation

The growing body of literature on the concept of value co-creation has generated considerable knowledge (Galvagno & Dalli, 2014), and it has been approached through various perspectives and models (Skaržauskaitė, 2013). These models share common trends and can be divided into three categories: Value co-creation as a service science; Value co-creation as technology and innovation development and Value co-creation as a form of collaborating between the customer and the business (Galvagno & Dalli, 2014).

For instance, one of the most prevalent models that views value co-creation as a form of collaborating between customers and businesses is the DART model (Galvagno & Dalli, 2014). Developed by Prahalad and Ramaswamy (2004), it entails as its main components Dialogue, Access, Risk-benefit and Transparency and proposes that these constitute the main elements of value co-creation between a firm and its customer (Leavy, 2012). While the DART model provides a solid basement for the value co-creation process, it needs additional layers to embrace other concepts within the service industry (Mazur and Zaborek, 2014). Alternatively, New Product Development (NPD) model often embraces value co-creation as a technology and innovation development (Galvagno & Dalli, 2014), by identifying four styles of value co-creation: Tinkering, Submitting, Co-designing and Collaborating (Agrawal & Rahman, 2015). Each of these styles suits a certain business situation in order to co-create valuable products and services for consumers (Vaisnore & Petraite, 2011). However, this model is challenged by the contradictions caused by the increasing complexity of customer interactions (Altun, Dereli and Baykasoğlu, 2013). With regards to of value co-creation as a service science, the Joint Sphere Model is based on the value-in-use concept, which refers to a value created while consuming and using products and services (Chathoth et al., 2016; Galvagno & Dalli, 2014). Value-in-use does not refer to the benefits obtained during the time of consuming the service only, but it extends further to include other dimensions of consumption (Agrawal & Rahman, 2013). These dimensions can be summarised into four types of value-in-use: physical usage (Grönroos, 2011), the mental usage (Payne et al., 2018), the virtual usage (Grönroos, 2011; Grönroos, Fisk and Sheth, 2013) and the possessive usage (Prebensen & Xie, 2017).

The above models, which attempt to explain the value co-creation process, are applied in different fields as well as in the tourism and service sectors (Galvagno & Dalli, 2014). This study adopts Grönroos's joint sphere model (2011) to develop a better understanding of how value can be co-created within the desert camps of Oman. Arguably, based on the fact that the value-in-use concept is designed to gain insights into customer experience and how it can be co-created (Prebensen & Xie, 2017), the joint sphere model enables service providers to capture different moments of the customers' consumption and shed light on the space in which value co-creation takes place (Grönroos et al., 2013).

As a result, the joint sphere model of value co-creation can provide firms with a better understanding of the service sector (Grönroos, 2011). In this study specifically, the model was adapted to incorporate the five key constructs of interest: marketing through the value in use concept (Grönroos, Fisk and Sheth, 2013); engagement (Eloranta & Matveinen, 2014), social media (Eloranta & Matveinen, 2014); authenticity (Counts, 2009) and place attachment (Suntikul & Jachna, 2016). The following diagram illustrates the conceptual framework of the study which was guided by the joint-sphere model of value co-creation (Figure 1).

Value-in-use

While viewed in an individual manner due to the customers' different perceptions of value (Vargo and Lusch, 2014), the concept of value-in-use is not isolated in customers' ecosystems, but is part of a dynamic system that links service providers and customers (Chandler & Vargo, 2011). Therefore, there is a need to establish a network that links customers' internal systems of creating value with the system of supplying the services (Chandler & Vargo, 2011; Vargo & Lusch, 2012). This network can be implemented through various methods including utilising social media platforms (Eloranta & Matveinen, 2014).

Since the role of hospitality organisations is to facilitate the creation of value through the process of obtaining feedback from their clients throughout the different phases of service consumption (Prebensen & Xie, 2017), opportunities of collaborating with customers should be identified and seized upon through marketing processes (Buhalis & Foerste, 2015). Businesses should adopt a marketing strategy which allows customers to re-design and customise the final output (Rozenes & Cohen, 2010). On this regard, value- in-use ensures a maximum level of confirming the customer's expectations after using certain products or services (Payne et al., 2018), and it extends further to create a memorable experience afterwards (Heinonen & Strandvik, 2009).

Value co-creation through social media use

Social media are platforms which enable users to generate content based on their experiences and perspectives (Albarran, 2013). In the tourism sectors especially, technology and social media platforms have changed aspects of the marketplace through marking a new role for the customer in the process of creating value (Albarran, 2013; Fotis et al., 2011; Gouillart & Ramaswamy, 2014; Yang et al., 2016). In today's marketplace, social media are viewed as tools to obtain feedback and enhance the quality of the company's services and products based on the user's role in feeding the evaluation process (Kaplan & Haenlein, 2010), and in articulating perceived values that can be captured to have a better understanding of their needs (Chesbrough, 2011). Customers' posts in social media platforms form a base for the ties and relationships between customers and organisations (Sorensen et al., 2017), and provide both parties with direct and accessible communication (Munar & Jacobsen, 2014). Therefore, technological interconnectivity has played a major role in improving and enhancing the tourist experience (Neuhofer et al., 2013).

When assessing the role of social media, it is also worth noting that tourists conduct many searches before travelling, therefore the communication circle starts with

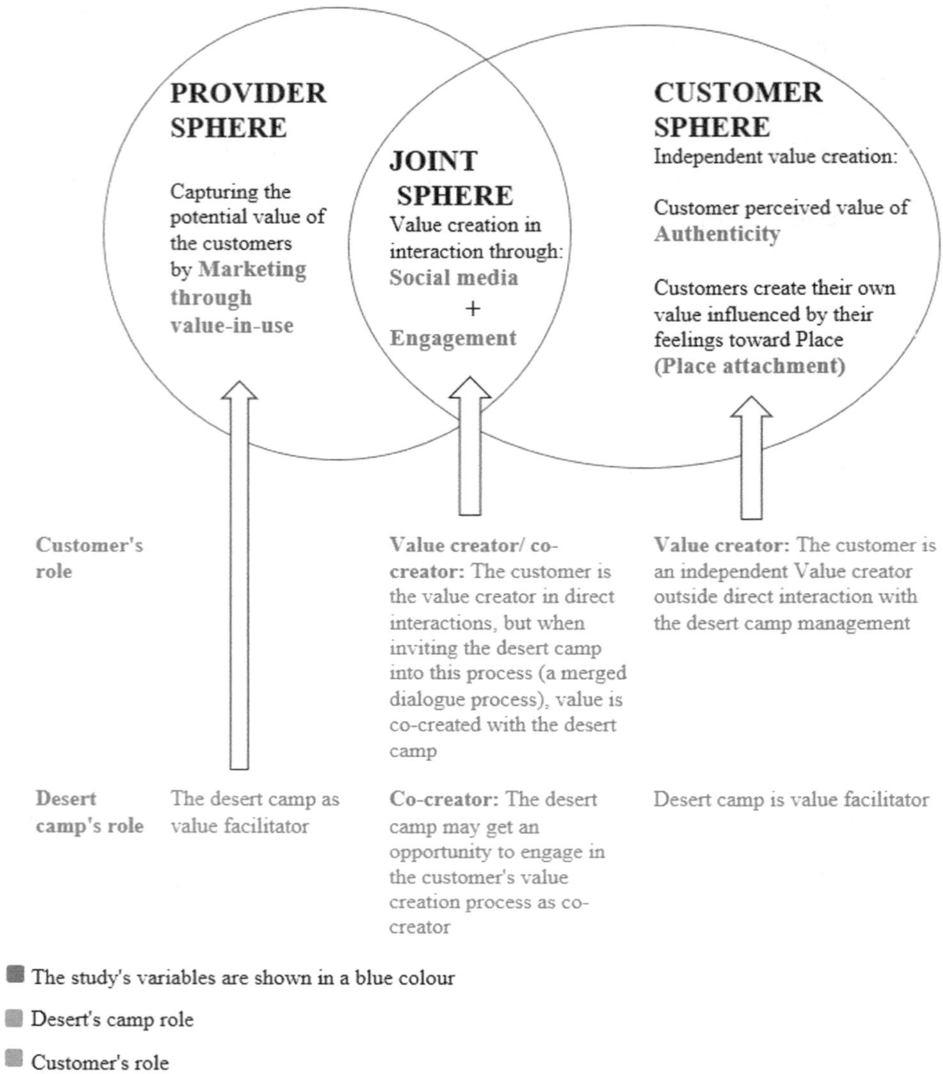

Figure 1. Conceptual framework.

customers revising other visitors' experiences before conducting their own experience. It then continues during the time of service and ends with the customers expressing their own views (Neuhofer et al., 2013). Hence, firms adopting social media as a tool for value co-creation should assess to what extent they understand their customers' requirements through increasing their level of engagement with them (Michopoulou & Moisa, 2018; Neuhofer et al., 2013).

Value co-creation through engagement

Engagement refers to the online and offline interactions between customers and organisations (Brodie et al., 2011). The recognition of the importance of the engagement process is based on the effective results obtained from the active participation of the customers through the value co-creation process (Brodie et al., 2011; Chathoth et al., 2016; Singh & Sonnenburg, 2012).

Gouillart and Ramaswamy (2014) state that value co-creation depends on the opportunities of interactions and engagement between customers and firms. Arguably, the development of technology and its accessibility through mobile devices has empowered customers to have an essential means of communication with the business (Luo et al., 2015). More precisely, the accessibility of information in the value co-creation process has become easier through the use of social media platforms (Gouillart & Ramaswamy, 2014). Therefore, businesses are investing on technology to increase customer engagement and to co-create value, as technology comes to be viewed as a tool to expand the range of the potential market in the short and in the long term (Eccleston et al., 2020; Gouillart & Ramaswamy, 2014).

Customer engagement does not only rely on their interactions with businesses, but it extends further to exclude the firm from the equation in some situations. This is known as customer to customer value co-creation (Rihova et al., 2018), where co-creation takes place between the customers in their own ecosystem. The visibility of these interactions becomes more obvious to service providers when customers share their experiences and opinions on social media platforms (Munar & Jacobsen, 2014), where the value outcomes formed through customer to customer co-creation contribute to shape a collective social value (Cheng, 2016).

Value co-creation through authenticity

Authenticity describes a perspective which is characterised by genuine and true aspects (Dieke et al., 2015). MacCannell (1973) argued that the continuous desire of customers to have a genuine experience has created a strong quest for authenticity in tourism and hospitality industries. In tourism studies, the literature shows different perceptions of the concept of authenticity (Dieke et al., 2015; Elomba & Yun, 2018) also evident and contested within Omani desert tourism (Chatty, 2016).

Object-based authenticity emerged from a museum background, where there is a need for experts to determine the originality of items and objects (Castéran & Roederer, 2013). This means that judgements of what is authentic are produced by experts based on a standard and static concept without involving tourists' opinions and experiences (Checa-Gismero, 2018; Steiner & Reisinger, 2006; Tiberghien, 2019). Such concept has been challenged by scholars like Kolar and Zabkar (2010) who argued that authenticity should be understood within its context and the perceptions of its receivers and their experiences. However, unlike certain objects' or touristic artefacts which authenticity can be contested (Steiner & Reisinger, 2006) the genuineness of the desert as a "pristine ecosystem and an authentic—unmodified, undisturbed and unstaged—natural space is unquestionable" (Moufakkir & Selmi, 2018, p. 114). Whilst object-based authenticity is applicable in desert tourism with regards to the physical environment (i.e sand dunes, clear skies, colours, stars) and objects and artefacts (i.e traditional linens and textiles), it is not adequate to describe the experiences within that space.

Consequently, the concept of constructive authenticity was introduced, where authenticity is formed by the development of the social perspectives of the customers (Olsen, 2002). This also disproves the static concept of authenticity mentioned above, as it suggests that authenticity is socially constructed over time (Olsen, 2002; Steiner & Reisinger, 2006). Wang (2010) has extended the idea of authenticity further by including the sphere of customer experience and arguing that in the hospitality industry authenticity is linked to the experiences of people, as confirmed by other studies (Castéran & Roederer, 2013; Checa-Gismero, 2018; Tiberghien, 2019). This is in alignment with the view of Counts (2009), who states that the authenticity of experiences provides opportunities for co-production as it allows the customers to be active co-creators of their own experiences. Wang also developed a concept to differentiate between the authenticity of objects and the authenticity of experience, which has led to the concept of existential authenticity (Wang, 2010). This concept suggests that as the perception of authenticity is based on experience, it is linked to intrapersonal sources depending on the tourists' feelings. The desert is a unique space (with its silence, unpredictable outlines, and raw nature) that is inextricably linked to existential authenticity as it enables contemplation and meditation because it highlights the finiteness of the human condition (Moufakkir & Selmi, 2018; Narayanan & Macbeth, 2009). The interactive multisensory experience also offers a "short-term break from the perceived inauthenticity of the everyday life" (Moufakkir & Selmi, 2018, p. 115). Hence, the perceived authenticity of desert camp experiences remains in the minds of the tourist, yet to be fully uncovered.

As managers encounter difficulty in harmonising the different and often contradictory types of authenticity (Kolar & Zabkar, 2010), businesses should adopt a strategy that originates from the customers' point of view of authenticity (Gilmore & Pine, 2015). However, in order to do so, managers should be aware of the main factors that affect the concept of authenticity in the modern market (Andriotis, 2011; Kolar & Zabkar, 2010).

Value co-creation and place attachment

The process of value co-creation in the tourism and hospitality industry is also related to the customers' relationship with the physical place they are visiting (Suntikul & Jachna, 2016). Mossberg (2007) argues that customers' experiences are linked to their emotional perspective of a place, as the value creation emerges through co-creating positive meanings and feelings between the customers and the visited place (Suntikul & Jachna, 2016). Based on this, Larsen (2007) argues that site engagement is of paramount importance in building up a memorable experience.

Place attachment can be divided in place dependence and place identity (Binkhorst & Den Dekker, 2009). Place dependence refers to the ability of the place to provide different services and to meet the needs and expectations of the customers (Larsen, 2007; Suntikul & Jachna, 2016). Place identity refers to devoting great attention to the customers' emotional ties and connection to the place (Binkhorst & Den Dekker, 2009). The place attachment element contributes in shaping all four areas of tourists' feelings and experiences: aesthetics, entertainment, education and escapism (Gilmore & Pine, 2015). Therefore, place attachment should be considered in value co-creation (Gilmore & Pine, 2015), as understanding customers' perceptions and feelings towards a place can help tourism organisations provide them with memorable experiences (Suntikul & Jachna, 2016). Another consideration on place attachment is the element of place social bonding (Scannell & Gifford, 2010; Vada et al., 2019) whereby tourists look at a destination favourably, due to the social interactions that take place at the destination (i.e. exchanging stories around the camp fire). Suntikul and Jachna (2016) argue that customer experience is co-created through place attachment when the service suppliers deal with this element as fundamental to value co-creation rather than as a service setting.

Building upon these considerations and responding to calls for further research on value co-creation within the tourism and hospitality sectors this study set out to understand key elements of value co-creation in the desert camps of Oman from both camp managers' and customers' perspectives.

Methodology

This research used a mixed methodology approach to provide the researchers the flexibility to investigate the topic with both quantitatively and qualitatively (Creswell, 2013; Tashakkori & Teddlie, 2016). The study deploys a convergent parallel design which entails collecting the qualitative and quantitative data simultaneously, analysing them separately, and then examining whether the findings complement each other by comparing the obtained results (Creswell and Plano Clark, 2006).

The role of the camp managers is inductively examined since there is limited knowledge about their role in the process of value co-creation (Mesoudi et al., 2018). Semi-structured interviews were conducted as they provide the interviewer and the interviewee with the flexibility to bring new ideas through a probing process (Bryman & Bell, 2015). The role of customers in the value co-creation process however has been reasonably examined in the literature (Im & Qu, 2017) and therefore here it is examined deductively and quantitatively, through the use of a survey. Questionnaire items were used from validated scales based on constructs adapted from the literature. Items for *place attachment* were adopted from Suntikul and Jachna (2016), *authenticity* from Kolar and Zabkar (2010), *engagement* from Braun et al. (2017), *social media* from Carlson, Rahman, Voola, and De Vries (2018) and *value-in-use* from Ranjan and Read (2014).

To collect data from the customers, at least one characteristic for the population was set (Saunders et al., 2016): having had an experience in desert camps of Oman. The camp managers were selected through purposive sampling, and the snowball method (Saunders et al., 2016) was then deployed to grow the sample size number by allowing the interviewees to nominate other eligible participants that could be involved in the study.

The surveys were processed and analysed through SPSS computer software (Field, 2014). The semi-structured interviews were analysed through thematic analysis so to identify, analyse, and report patterns through positioning the information into categories within data (Braun & Clarke, 2006; Bryman & Bell, 2015). Finally, findings from quantitative and qualitative data were converged and used to answer the study main objectives.

Findings and discussion

Analysis of qualitative data

There are 20 camps in Oman in total (Alshaibi, 2018), and 8 camp managers were interviewed. The researcher was successful in reaching 40% of the whole population of the desert camp managers. The data saturation was reached in the fifth interview, however, the researcher carried out additional interviews to ensure robustness of data. The interviews were semi-structured which allowed exploring and grouping the data gathered into specific themes. Based on the literature review on value co-creation, the following themes that consist of other sub themes emerged through the process of the data analysis: authenticity, engagement, place attachment, and social media use (Figure 2).

With regards to the central concept of value in use, findings revealed how camp managers use the joint sphere model with their customers to co-create value. In particular, the lack of adopting a comprehensive strategy to implement the different aspects of the value-in-use model was identified. In fact, interviewees highlighted that only 2 out of the 4 dimensions (Grönroos, Fisk and Sheth, 2013) relating to the value in use concept are being considered: the physical usage and the possessive usage. The other two dimensions (mental usage and virtual usage), which require adopting mechanisms to engage directly with the customers, received less attention from the interviewees. They believed that the joint sphere of value co-creation with their customers takes place during the guests' stay only: "Our service design is a result of discussions with the tourists who have visited us. We take their discussions seriously. That's why we have one of the best services in the desert camps of Oman" (Participant 1). The lack of establishing direct engagement links with customers prior to and after their arrival therefore limits the ability of the desert camp managers to create value through using mental usage and virtual usage in the value-in-use concept.

The sub themes of authenticity, engagement, place attachment, and social media use are discussed later on in this paper in conjunction with the quantitative data.

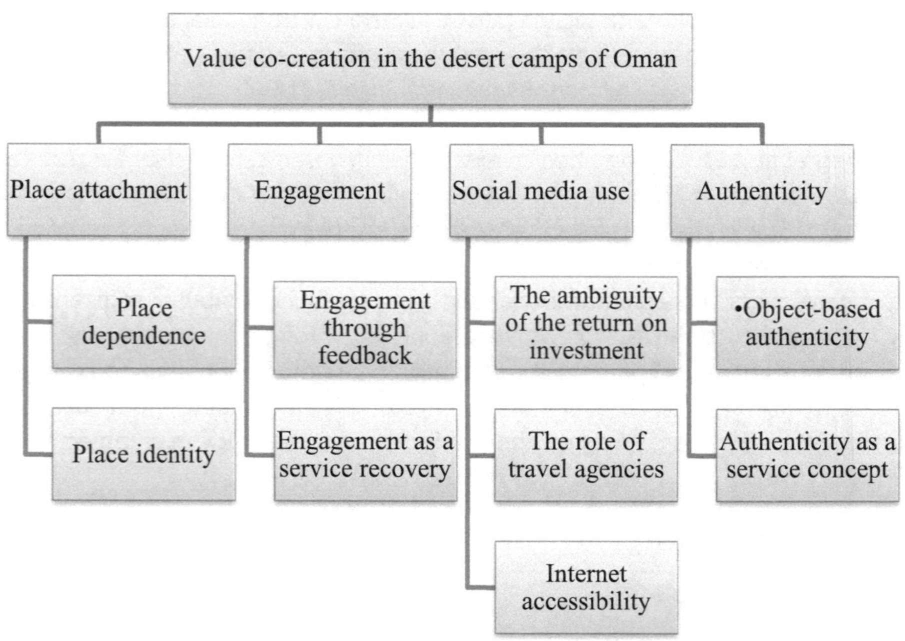

Figure 2. Thematic map.

Analysis of quantitative data

The researcher was successful in collecting 336 questionnaires on the topic of value co-creation in the desert camps of Oman. Of these 336 questionnaires, 324 were suitable for data analysis. The data collected from the desert camp customers accommodate three demographic details: age, gender and origin. In terms of gender, 57.7% of the sample were male and 42.3%, female. Most of the participants were from the age group of 40–65 years, which presented a percentage of 46%. Other age groups of 18–25, 25–40 and 65+ were 12%, 15% and 27%, respectively. The participants' background varied between 19 nationalities. The vast majority came from Germany, France, UK and Italy.

Results showed that for each indicator standardised loadings were acceptable (from 0.578 to 0.895). The Cronbach's alpha value is an indicator to examine the internal consistency of each construct and should be at least 0.70 to indicate high internal construct consistency. Table 1 shows the Cronbach's alpha values for all constructs ranged from 0.713 to 0.815. Results for composite reliability (CR) were over 0.70 and ranged from 0.78 to 0.86, indicating internal consistency of the constructs. For convergent validity to be considered good, the average variance extract (AVE) values should be greater than 0.50 (Andersson, Forsgren, & Holm, 2001); which was true for all the constructs in this study.

Regression results indicate that Place Attachment ($\beta = 0.22$; $p < 0.05$), Authenticity ($\beta = 0.16$; $p < 0.05$), Customer Engagement and ($\beta = 0.57$; $p < 0.001$) were all good predictors of Value in Use, however, social media was not significant. The overall explanatory power of the model was 62% (Adj R2= .625) (Figure 3).

Customer Engagement proved to be the strongest predictor of the proposed constructs accounting for over 50% ($\beta = 0.57$; $p < 0.001$) of the variance of Value in Use. This finding suggests that customers want to be actively involved and participate in value creation for both themselves and the company. This reinforces the notion that customers' ability to influence the value creation process so that it meets their needs, wants and expectations is linked to loyalty and satisfaction (Mathis et al., 2016). Place Attachment and Authenticity clearly linked to Value in Use, however the relationships observed were weak ($\beta = 0.22$ and $\beta = 0.16$ respectively). Surprisingly, there was no relationship between social media and value in use. This could be explained by the fact that there is no internet connectivity at the desert camps and the social media presence from the camp perspective is minimal.

The next section discusses the key constructs examined in this study from both customer and supplier perspectives by merging qualitative and quantitative data.

Customer engagement

The empirical data obtained from the customer survey show that engagement is by far the strongest predictor of value than any other construct. This is in alignment with the literature which shows that engagement is important in the active participation of customers in the value co-creation process and can take different forms including online and offline activities (Brodie et al., 2011; Hill & Steemers, 2017)

However, data from the camp managers show that there is a tendency to engage with customers through offline practices during the guests' stay to obtain their feedback and

Table 1. Constructs, item loadings, CA, Mean and SD.

Variables with corresponding measurement items	Cronbach alpha	Item loadings	Mean	SD
Value in use	.815			
The desert camp experience is a memorable experience		.684	4.15	1.38
Depending upon the nature of my own participation in my experience in the desert camp, my experiences might be different from other customers		.692	4.03	1.45
It was possible for me to improve my experience through experimenting and trying new activities in the desert camp		.785	3.93	1.39
Social media use	.801			
The social media platforms of the desert camp enhance my knowledge about the brand		.621	4.19	1.34
The social media platforms of the desert camp enhance my knowledge about the camp's offerings		.842	4.15	1.47
The social media platforms of the desert camp help me to obtain solutions to specific brand related problems that I have		.814	4.34	1.19
The users of the social media platforms of the desert camp form an entity		.895	3.96	1.23
The users of the social media platforms of the desert camp have a bond		.587	4.40	1.64
The users of the social media platforms of the desert camp have goals in common		.905	4.65	1.37
The social media platforms of the desert camp are fun		.818	4.54	1.52
The social media platforms of the desert camp are exciting		.859	4.31	1.29
The social media platforms of the desert camp are entertaining		.860	4.46	1.46
Customer Engagement	.752			
When I am engaging myself with the desert camp, I feel strong and vigorous		.878	4.40	1.24
I can continue engaging myself with the desert camp for very long periods of time		.745	4.65	1.27
I keep on engaging myself with the desert camp even when things do not go well		.798	4.54	1.32
Authenticity	.713			
The overall architecture and impression of the desert camp inspired me.		.675	4.55	1.58
I liked the peculiarities about the interior design/furnishings.		.813	4.58	1.34
I liked the way the desert camp blends with the attractive landscape/ scenery/historical ensemble/town, which offers many other interesting places for sightseeing		.765	4.43	1.71
I liked special arrangements, events, concerts, celebrations connected to the desert camp		.724	4.45	1.53
This visit provided a thorough insight into the Omani historical era		.675	4.35	1.48
I liked the calm and peaceful atmosphere during the visit.		.654	4.28	1.51
I felt connected with human history and civilisation		.876	3.78	1.71
During the visit I felt the related history, legends and historical personalities.		.697	3.85	1.67
Place attachment	.794			
The atmosphere of the desert camp gives a feeling of the unique character of the desert		.651	4.23	1.51
The desert camp is a comfortable and pleasant place to be		.578	4.16	1.37
The place in which the camp is located is promoted as a must-see place for visitors to Oman		.536	4.65	1.57
I would personally recommend this site as a must-see place to visitors to Oman		.679	4.11	1.62
I will remember this site long after leaving Oman		.826	4.52	1.42
I would like to come back to this site if I could		.632	4.19	1.57

identify improvement opportunities "Customer engagement is to get feedback from them during their stay and act positively on these feedbacks to gain their satisfaction" (Participant 6); whereas online engagement is not practiced due to the limited implementation of social media.

Interviewees also highlighted the role of travel agencies in the desert camps of Oman to engage, attract and communicate with the guests before and after their visits. In this

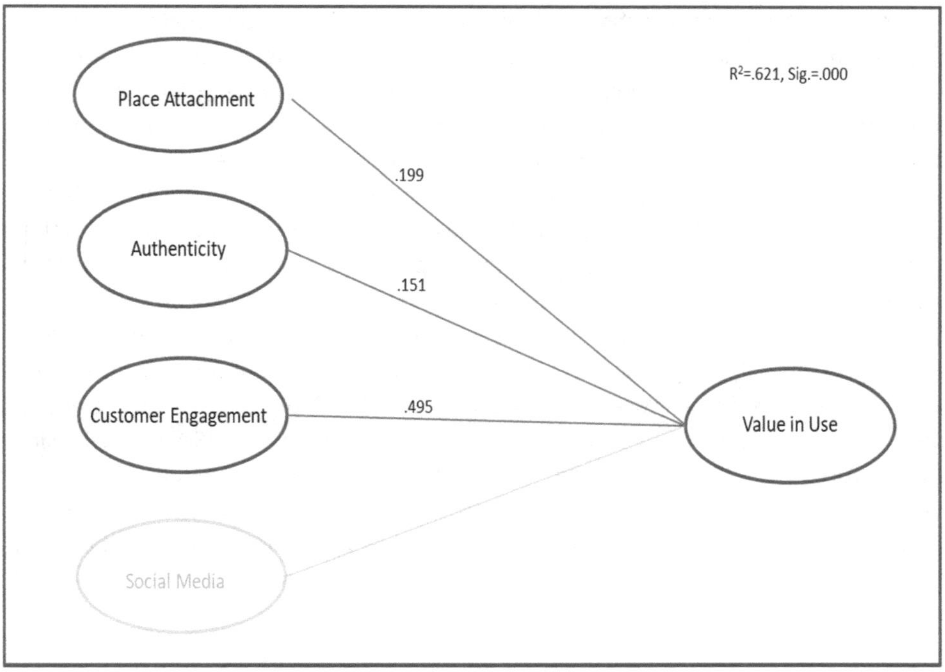

Figure 3. Model output.

regard, studies discuss that while on-the-spot engagement with customers contributes to obtaining quicker feedback, it has many limitations in providing the guests with their desired experience, if solely used (Fotis et al., 2011; Hill & Steemers, 2017). Engagement with customers should be practised during the different phases of the guest journey, which consists of trip planning, actual consumption of the service and communication with the guests after their departure (Fotis et al., 2011). It could therefore be presumed that engagement is important for customers in value creation but for most of their "journey" phases it occurs via intermediaries rather than the camps directly. Hence, intermediaries fill the engagement gap allowed by the camps.

Authenticity

The results from customer survey demonstrated that the authenticity factor is important in predicting value in use. This finding supports the premise put forward by Prebensen et al. (2016) that both object-based authenticity and existential authenticity play a role in value co-creation. The low score however may be due to the fact that the measured construct entailed items for both object-based and existential authenticity. Dieke et al. (2015) argue that existential authenticity in tourism activities is linked to its nature of being activity-centric, whereas object-based authenticity focuses on objects only. Hence, if customers are more focused on activity, then the scores of existential authenticity would be high, whilst scores for object-based authenticity low.

The qualitative data obtained from the desert camp managers also confirmed that authenticity is considered to be the core element in creating values for the guests: "We provide our guests with the values that they are looking for, through providing them with authentic Omani desert experience" (Participant 7) and "Providing the guests with authentic experience is key for success. Therefore, our main theme in the camp is an atmosphere of authenticity" (Participant 5). This is in alignment with the Ministry of Tourism's target to provide customers with an "Authentic Arabic experience". Interviewees viewed that desert tourism consists of two main components: cultural and natural. The cultural component is associated with the customers' quest for authenticity and their curiosity to discover and explore the local culture, whereas the natural component is associated with the customers' appreciation of the landscape and their desires to perform certain activities.

The implementation of authenticity as a focal point of managers' service concept design is practised through creating a servicescape which comprises local elements, local furniture, local décor touches and local handicrafts. Arguably, this tacit creation of value through authenticity is based on object-based authenticity (Steiner & Reisinger, 2006; Tiberghien, 2019), and it doesn't factor other emerging notions of authenticity in tourism. Existential authenticity (Wang, 2010) for example, in which the customers should be highly involved in processes of creating values through merging their understanding of authenticity, is not widely understood nor implemented.

Place attachment

The role of place attachment in the process of value co-creation in the desert camps of Oman has here been split in two components, place identity (Binkhorst & Den Dekker, 2009) and place dependence (Suntikul & Jachna, 2016), as these represent the two main practices deployed by the desert camp managers in Oman to ensure value co-creation through place attachment. However, customers' view of place attachment does not seem to be strong in predicting value in use.

Co-creating value with the customers through using the relationship between place identity and authenticity is a strategy widely used by the camp managers

> The location of our camp was chosen in respect to our traditions and the Omani architectural style. We inform our customers that our place was chosen with these considerations. Customers therefore develop curiosity to ask different questions about our life and appreciate our authentic concept at the camp to which they get emotionally attached to this place. (Participant 1)

The place identity element was therefore found to have a great impact on the process of co-creating value in the desert camps of Oman. This finding is in alignment with Binkhorst and Den Dekker (2009)'s argument, which states that place identity is connected to the guests' emotions and it has a greater influence on their perceived values. However, other interviewees viewed place identity as a mean to shape the guest's experience through presenting services and activities appropriate to the camp. They believe that guests develop a strong connection to the camp's place based on emotional ties that could be developed through activities. In this regard, participant 7 stated: "As you can see, there is nothing around here except the sands, but we bring life to this place through the camp activities where the guests can have a memorable experience".

Regarding place dependence (Suntikul & Jachna, 2016), the managers of the camps which are located deep in the desert claimed that their locations provided the customers with a real, authentic desert experience as it responds to their quest for escaping from the stress of the cities

> Being deep into the desert where you are just surrounded with the beautiful sand dunes, gives you unique feelings and thoughts. We are located far away from the city where our guests can free their minds and have great level of tranquillity, (Participant 4)

On the other hand, the camps which are located near the cities claimed that their locations have advantages over the other camps

> Our location is very accessible, and our guests can reach us at ease unlike the other camps. Our proximity to the main roads facilitates obtaining fresh and daily supplies. The guests appreciate this top quality restaurant service and they get attached to the place of our camp. (Participant 2)

Overall, the camp managers stated that the desert camps rely heavily on the surrounding landscape to create an experience for their customers. However, a review of Gilmore and Pine's study (2015) on the multi-dimensional aspects that develop the customers' attachment to the place (aesthetic, educational, escapism and entertainment) indicates that camp managers focus mainly on two elements: escapism and entertainment; meaning that service providers are failing to use place attachment as a fundamental element in value co-creation (Gilmore & Pine, 2015; Suntikul & Jachna, 2016). It can be concluded that through considering these four factors, the camp managers could gain deeper insights into using place attachment to co-create values with their customers. This may also explain why from the customers' perspective the importance of place attachment in predicting value was limited.

Social media use

Results from the customer survey yielded a rather surprising finding; that there is no significant relationship between social media and value in use. Hence, customers do not consider the use of social media as an important factor as a means for interaction with service providers to facilitate value co-creation. This is comes in stark contrast to Neuhofer et al. (2013) who argue that social media use has changed aspects of the marketplace through marking a new role for the customers as co-creators. However, this finding seems to be context specific as explained by the qualitative data, which explored the role of the camp managers in using social media. Interviewees were hesitant to use social media platforms to co-create value with the guests: while some do not have any presence in social media at all, others use the platforms only as means of promotion. Three main barriers to the use of social media where identified: the role of travel agencies as mediators, the ambiguity of the return on investment and internet accessibility.

It can be argued that, since desert camp experiences are designed, marketed and organised by travel agencies, the value co-creation through social media takes place between the customers and the travel agencies. Sharda and Chatterjee (2011) claim that the presence of intermediaries disconnects the customers from interacting directly with the service providers and Cannas (2018) expands on this argument highlighting the risk for unequal power imbalances between the actors involved in the value co-creation process. Therefore, it is recommended that the camps use their own social media platforms to facilitate the process of value co-creation through building direct bridges with their customers.

Apprehension for the return on the investment also prevents the camp mangers from developing their social media platforms to increase the level of the engagement with their customers. The participants showed difficulties in assessing the real influences of social media, and showed reluctance towards investing on a social media strategy, as stated by participant 2: "We focus on our service level and we have good reviews which market our camp. Investing on social media requires financial resources. We should instead focus on our service".

Most participants highlighted the issue of technology limitations in accessing social media platforms, as most of the camps are located in remote places. In addition, most desert camps do not provide WIFI services, and this seems to add value to the guests' experience "In the desert, our customers enjoy our concept of getting themselves involved in different and adventurous lifestyles where they disconnect themselves from the phone addiction" (Participant 5). The researchers enquired about using the city offices to communicate with the customers on social media platforms, but the participants showed disinclination due to the limited resources, roles and numbers of employees "We have one employee working at the office, he can't do everything" (Participant 4).

Hence, it seems that customers do not place any importance on social media in creating value at the desert camps because either there is no option for interaction altogether (camps with no social media presence) or the accounts that exist do not interact but are only used for push messages and advertising. It is therefore important to consider the camps' barriers in using social media and develop solutions that will enable value creation.

Conclusion

This research aimed to explore the various elements and factors that can contribute to the process of value co-creation in the desert camps of Oman. The concept of value co-creation was investigated through the perspectives of the customers as well as those of the camp managers. This investigation revealed how these two parties co-create value and highlighted the discrepancies between them. Therefore, this study adds to the debate on identifying the common sphere between customers and service providers and defines this sphere within the context of service to co-create value in desert camps. Hence, this study contributes to knowledge by deepening our understanding of the process of value co-creation within unique destinations, in this case desert camps. It also establishes factors that are entailed and can enhance value in use. The study showed that authenticity, place attachment and engagement were strong predictors of value in use whilst social media was not.

In particular, customers' quest for authenticity entailed both existential and object-based authenticity, whilst camp managers responded to their customers' need through object-based authenticity only. The desert camp managers engaged with their customers mostly offline and during the guest's stay, but customers want to engage with the desert camps both online and offline. Customers consider place identity as an element which has a greater influence in value co-creation than place dependence. Desert camp managers are therefore co-creating value with customers through using the relationship between place identity and authenticity and by using the place identity element to design the camp's activities. With regards to social media it was evident that lack of skill and knowledge for the camp managers' part, absence of internet access at location and a dominant and even interfering role of intermediaries, made the option of co-creating value with customers through social media unattainable.

This study has implications for practitioners. Specifically, it is important that camp managers and managers of other unique and remote locations focus on engaging with the guests before, during and after their arrival. This assists in enhancing their experience and in co-creating shared value between the destinations and their guests. This can be achieved through merging the offline and the online engagement into a single component which represents the guests' requirements during the different phases of their journey. Managers should also concentrate on the value-in-use concept and all of its four dimensions (including mental usage and virtual value usage) in order to address customers' needs and preferences more effectively and co-create value for all parties involved.

Disclosure statement

No potential conflict of interest was reported by the author(s).

References

Agrawal, A., & Rahman, Z. (2015). Roles and resource contributions of customers in value co-creation. *International Strategic Management Review*, *3*(1-2), 144–160. https://doi.org/10.1016/j.ism.2015.03.001

Albarran, A. (2013). *The social media industries*. Routledge.

Allan, M. (2016). Place attachment and tourist experience in the context of desert tourism – the case of Wadi Rum. *Czech Journal of Tourism*, *5*(1), 35–52. https://doi.org/10.1515/cjot-2016-0003

Alshaibi, T. (2018). Oman camps. [online] *Oman Daily*. http://www.omandaily.om/ HYPERLINK "http://www.omandaily.om/631970/"631970 HYPERLINK "http://www.omandaily.om/631970/"/

Altun, K., Dereli, T., & Baykasoğlu, A. (2013). Development of a framework for customer co-creation in NPD through multi-issue negotiation with issue trade-offs. *Expert Systems with Applications*, *40*(3), 873–880.

Andersson, U., Forsgren, M., & Holm, U. (2001). Subsidiary embeddedness and competence development in MNCs a multi-level analysis. *Organization Studies*, *Vol. 22*(6), 1013–1034.

Andriotis, K. (2011). Genres of heritage authenticity. *Annals of Tourism Research*, *38*(4), 1613–1633. https://doi.org/10.1016/j.annals.2011.03.001

Baker, M., & Mearns, K. (2017). Applying sustainable tourism indicators to measure the sustainability performance of two tourism lodges in the Namib desert. *African Journal of Hospitality, Tourism and Leisure*, *6*(2), 1–22.

Bimonte, S., & Punzo, L. (2016). Tourist development and host–guest interaction: An economic exchange theory. *Annals of Tourism Research*, *58*, 128–139. https://doi.org/10.1016/j.annals.2016.03.004

Binkhorst, E., & Den Dekker, T. (2009). Agenda for co-creation tourism experience research. *Journal of Hospitality Marketing & Management*, *18*(2-3), 311–327. https://doi.org/10.1080/19368620802594193

Boksberger, P., & Melsen, L. (2011). Perceived value: A critical examination of definitions, concepts and measures for the service industry. *Journal of Services Marketing*, *25*(3), 229–240. https://doi.org/10.1108/08876041111129209

Braun, C., Hadwich, K., & Bruhn, M. (2017). How do different types of customer engagement affect important relationship marketing outcomes? An empirical analysis. *Journal of Customer Behaviour*, *16*(2), 111–144. https://doi.org/10.1362/147539217X14909732699525

Braun, V., & Clarke, V. (2006). Using thematic analysis in psychology. *Qualitative Research in Psychology*, *3*(2), 77–101. https://doi.org/10.1191/1478088706qp063oa

Brodie, R., Hollebeek, L., Jurić, B., & Ilić, A. (2011). Customer engagement. *Journal of Service Research*, *14*(3), 252–271. https://doi.org/10.1177/1094670511411703

Bryman, A., & Bell, E. (2015). *Business research methods*. Oxford University Press.

Buhalis, D., & Foerste, M. (2015). Socomo marketing for travel and tourism: Empowering co-creation of value. *Journal of Destination Marketing & Management*, *4*(3), 151–161. https://doi.org/10.1016/j.jdmm.2015.04.001

Cannas, R. (2018). Diverse economies of collective value co-creation: The open monuments event. *Tourism Planning & Development*, *15*(5), 535–550. https://doi.org/10.1080/21568316.2018.1505651

Carlson, J., Rahman, M., Voola, R., & De Vries, N. (2018). Customer engagement behaviours in social media: Capturing innovation opportunities. *Journal of Services Marketing*, *32*(1), 83–94. https://doi.org/10.1108/JSM-02-2017-0059

Castéran, H., & Roederer, C. (2013). Does authenticity really affect behavior? The case of the Strasbourg Christmas market. *Tourism Management*, *36*, 153–163. https://doi.org/10.1016/j.tourman.2012.11.012

Chandler, J., & Vargo, S. (2011). Contextualization and value-in-context: How context frames exchange. *Marketing Theory*, *11*(1), 35–49. https://doi.org/10.1177/1470593110393713

Chathoth, P., Ungson, G., Harrington, R., & Chan, E. (2016). Co-creation and higher order customer engagement in hospitality and tourism services. *International Journal of Contemporary Hospitality Management*, *28*(2), 222–245. https://doi.org/10.1108/IJCHM-10-2014-0526

Chatty, D. (2016). Heritage policies, tourism and pastoral groups in the sultanate of Oman. *Nomadic Peoples*, *20*(2), 200–215. https://doi.org/10.3197/np.2016.200203

Checa-Gismero, P. (2018). Global contemporary art tourism: Engaging with Cuban authenticity through the Bienal de La Habana. *Tourism Planning & Development*, *15*(3), 313–328. https://doi.org/10.1080/21568316.2017.1399435

Cheng, M. (2016). Sharing economy: A review and agenda for future research. *International Journal of Hospitality Management*, *57*, 60–70. https://doi.org/10.1016/j.ijhm.2016.06.003

Chesbrough, H. (2010). Business model innovation: Opportunities and barriers. *Long Range Planning*, *43*(2-3), 354–363. https://doi.org/10.1016/j.lrp.2009.07.010

Chesbrough, H. (2011). The case for open services innovation: The Commodity Trap. *California Management Review*, *53*(3), 5–20. https://doi.org/10.1016/j.lrp.2009.07.010

Cooke, B. (2010). Tourism in deserts. *Geography Review*. (09507035)

Counts, C. (2009). Spectacular design in museum Exhibitions. *Curator: The Museum Journal*, *52*(3), 273–288. https://doi.org/10.1111/j.2151-6952.2009.tb00351.x

Creswell, J. (2013). *Research design*. SAGE.

Creswell, J., & Plano Clark, V. (2006). *Designing and conducting mixed methods research*.

Dieke, P., Heitmann, S., & Robinson, P. (2015). *Research themes for tourism*. CABI.

Eccleston, R., Hardy, A., & Hyslop, S. (2020). Unlocking the potential of tracking technology for co-created tourism planning and development insights from the tourism Tracer Tasmania Project. *Tourism Planning & Development*, *17*(1), 82–95. https://doi.org/10.1080/21568316.2019.1683884

Elomba, M., & Yun, H. (2018). Souvenir authenticity: The perspectives of local and Foreign tourists. *Tourism Planning & Development*, *15*(2), 103–117. https://doi.org/10.1080/21568316.2017.1303537

Eloranta, V., & Matveinen, J. (2014). Accessing value-in-use information by integrating social platforms into service offerings. *Technology Innovation Management Review*, *4*(4), 26–34. https://doi.org/10.22215/timreview/782

Field, A. (2014). *Discovering statistics using ibm spss statistics + spss version 22.0*. Sage Publications.

Fotis, J., Buhalis, D., & Rossides, N. (2011). Social media impact on Holiday travel planning. *International Journal of Online Marketing*, *1*(4), 1–19. https://doi.org/10.4018/ijom.2011100101

Franke, N., & Schreier, M. (2010). Why customers value self-designed products: The importance of process Effort and Enjoyment*. *Journal of Product Innovation Management*, *27*(7), 1020–1031. https://doi.org/10.1111/j.1540-5885.2010.00768.x

FSG. (2018). *Oman Tourism Report.* [online] English. https://store.fitchsolutions.com/oman-tourism-report.html

Galvagno, M., & Dalli, D. (2014). Theory of value co-creation: A systematic literature review. *Managing Service Quality*, *24*(6), 643–683. https://doi.org/10.1108/MSQ-09-2013-0187

Gilmore, J., & Pine, B. (2015). *Authenticity*. Harvard Business Review Press.

Gouillart, F., & Ramaswamy, V. (2014). *The power of co-creation*. Free Press.

Grayson, K., & Martinec, R. (2004). Consumer perceptions of iconicity and indexicality and their influence on assessments of authentic market offerings. *Journal of Consumer Research*, *31*(2), 296–312. https://doi.org/10.1086/422109

Grönroos, C. (2011). A service perspective on business relationships: The value creation, interaction and marketing interface. *Industrial Marketing Management*, *40*(2), 240–247. https://doi.org/10.1016/j.indmarman.2010.06.036

Grönroos, C., Fisk, R., & Sheth, J. (2013). *Service marketing*. SAGE Publications.

Heinonen, K., & Strandvik, T. (2009). Monitoring value-in-use of e-service. *Journal of Service Management*, *20*(1), 33–51. https://doi.org/10.1108/09564230910936841

Henderson, G. C. (2015). The development of tourist destinations in the Gulf: Oman and Qatar compared. *Tourism Planning & Development*, *12*(3), 350–361. https://doi.org/10.1080/21568316.2014.947439

Hill, A., & Steemers, J. (2017). Media industries and engagement. *Media Industries Journal*, *4*(1). https://doi.org/10.3998/mij.15031809.0004.105

Hsiao, C., Lee, Y., & Chen, W. (2015). The effect of servant leadership on customer value co-creation: A cross-level analysis of key mediating roles. *Tourism Management*, *49*, 45–57. https://doi.org/10.1016/j.tourman.2015.02.012

Im, J., & Qu, H. (2017). Drivers and resources of customer co-creation: A scenario-based case in the restaurant industry. *International Journal of Hospitality Management*, *64*, 31–40. https://doi.org/10.1016/j.ijhm.2017.03.007

Jangra, R., & Kaushik, S. P. (2018). Analysis of trends and seasonality in the tourism industry: The case of a cold desert destination- Kinnaur, Himachal Pradesh. *African Journal of Hospitality, Tourism and Leisure*, *7*(1), 1–16.

Kaplan, A., & Haenlein, M. (2010). Users of the world, unite! The challenges and opportunities of social media. *Business Horizons*, *53*(1), 59–68. https://doi.org/10.1016/j.bushor.2009.09.003

Ketonen-Oksi, S., Jussila, J., & Kärkkäinen, H. (2016). Social media based value creation and business models. *Industrial Management & Data Systems*, *116*(8), 1820–1838. https://doi.org/10.1108/IMDS-05-2015-0199

Kolar, T., & Zabkar, V. (2010). A consumer-based model of authenticity: An oxymoron or the foundation of cultural heritage marketing? *Tourism Management*, *31*(5), 652–664. https://doi.org/10.1016/j.tourman.2009.07.010

Larsen, S. (2007). Aspects of a psychology of the tourist experience. *Scandinavian Journal of Hospitality and Tourism*, *7*(1), 7–18. https://doi.org/10.1080/15022250701226014

Leavy, B. (2012). Collaborative innovation as the new imperative – design thinking, value co-creation and the power of "pull". *Strategy & Leadership*, *40*(2), 25–34. https://doi.org/10.1108/10878571211209323

Lugosi, P. (2014). Mobilising identity and culture in experience co-creation and venue operation. *Tourism Management*, *40*, 165–179. https://doi.org/10.1016/j.tourman.2013.06.005

Luo, N., Zhang, M., & Liu, W. (2015). The effects of value co-creation practices on building harmonious brand community and achieving brand loyalty on social media in China. *Computers in Human Behavior*, *48*, 492–499. https://doi.org/10.1016/j.chb.2015.02.020

MacCannell, D. (1973). Staged authenticity arrangements of social space in tourist Settings. *American Journal of Sociology*, *79*(3), 589–603. https://doi.org/10.1086/225585

Mathis, E., Kim, H., Uysal, M., Sirgy, J., & Prebensen, N. (2016). The effect of co-creation experience on outcome variable. *Annals of Tourism Research*, *57*, 62–75. https://doi.org/10.1016/j.annals.2015.11.023

Mazur, J., & Zaborek, P. (2014). Validating dart model. *International Journal of Management and Economics*, *44*(1), 106–125.

Mesoudi, A., Brookes, T., & Fitch, C. (2018). *Oman: The Sea of Sand and Mists – Geographical*. [online] Geographical.co.uk. http://geographical.co.uk/expeditions/item/ HYPERLINK "http://geographical.co.uk/expeditions/item/2479-the-sea-of-sand-and-mists"2479 HYPERLINK "http://geographical.co.uk/expeditions/item/2479-the-sea-of-sand-and-mists"-the-sea-of-sand-and-mists

Michopoulou, E., & Moisa, D. (2018). Hotel social media metrics: The ROI dilemma. *International Journal of Hospitality Management*, *76*(Part A), 308–315. https://doi.org/10.1016/j.ijhm.2018.05.019

Mossberg, L. (2007). A marketing approach to the tourist experience. *Scandinavian Journal of Hospitality and Tourism*, *7*(1), 59–74. https://doi.org/10.1080/15022250701231915

Moufakkir, O., & Selmi, N. (2018). Examining the spirituality of spiritual tourists: A Sahara desert experience. *Annals of Tourism Research*, *70*, 108–119. https://doi.org/10.1016/j.annals.2017.09.003

MTO. (2018). *Oman*. [online] Omantourism.gov.om. https://omantourism.gov.om/wps/portal/mot/tourism/oman/home/sultanate/regions/asharqiyah

Munar, A., & Jacobsen, J. (2014). Motivations for sharing tourism experiences through social media. *Tourism Management*, *43*, 46–54. https://doi.org/10.1016/j.tourman.2014.01.012

Narayanan, Y., & Macbeth, J. (2009). Deep in the desert: Merging the desert and the spiritual through 4WD tourism. *Tourism Geographies*, *11*(3), 369–389. https://doi.org/10.1080/14616680903032783

Neuhofer, B., Buhalis, D., & Ladkin, A. (2013). A typology of technology-enhanced tourism experiences. *International Journal of Tourism Research*, *16*(4), 340–350. https://doi.org/10.1002/jtr.1958

Olsen, K. (2002). Authenticity as a concept in tourism research. *Tourist Studies*, *2*(2), 159–182. https://doi.org/10.1177/146879702761936644

Pappas, N., & Michopoulou, E. (2019). Hospitality in a changing world. *Hospitality & Society*, *9*(3), 259–265. https://doi.org/10.1386/hosp_00001_2

Payne, A., Ulaga, W., Frow, P., & Eggert, A. (2018). Conceptualizing and communicating value in business markets: From value in exchange to value in use. *Industrial Marketing Management*, *69*, 80–90. https://doi.org/10.1016/j.indmarman.2018.01.018

Pérez-Liu, R., & Tejada-Tejada, M. (2020). Citizens' view of their relationship with tourism in a desert coastal area. *Journal of Coastal Research*, *95*(sp1), 925-929. https://doi.org/10.2112/SI95-180.1

Prahalad, C., & Ramaswamy, V. (2004). *The future of competition*. Harvard Business Review Press.

Prebensen, N., Chen, J., & Uysal, M. (2016). *Creating experience value in tourism*.

Prebensen, N., & Xie, J. (2017). Efficacy of co-creation and mastering on perceived value and satisfaction in tourists' consumption. *Tourism Management*, *60*, 166–176. https://doi.org/10.1016/j.tourman.2016.12.001

Ranjan, K., & Read, S. (2014). Value co-creation: Concept and measurement. *Journal of the Academy of Marketing Science*, *44*(3), 290–315. https://doi.org/10.1007/s11747-014-0397-2

Rihova, I., Buhalis, D., Gouthro, M., & Moital, M. (2018). Customer-to-customer co-creation practices in tourism: Lessons from customer-dominant logic. *Tourism Management*, *67*, 362–375. https://doi.org/10.1016/j.tourman.2018.02.010

Rozenes, S., & Cohen, Y. (2010). *Handbook of research on strategic alliances and value co-creation in the service industry*.

Saunders, M., Lewis, P., & Thornhill, A. (2016). *Research methods for business students*.

Scannell, L., & Gifford, R. (2010). Defining place attachment: A tripartite organizing framework. *Journal of Environmental Psychology*, *30*(1), 1–10. https://doi.org/10.1016/j.jenvp.2009.09.006

Sharda, K., & Chatterjee, L. (2011). Configurations of outsourcing firms and organizational performance. *Strategic Outsourcing: An International Journal*, *4*(2), 152–178. https://doi.org/10.1108/17538291111147991

Singh, S., & Sonnenburg, S. (2012). Brand Performances in social media. *Journal of Interactive Marketing*, *26*(4), 189–197. https://doi.org/10.1016/j.intmar.2012.04.001

Skaržauskaitė, M. (2013). Measuring and managing value co-creation process: Overview of existing Theoretical models. *Social Technologies*, *3*(1), 115–129. https://doi.org/10.13165/ST-13-3-1-08

Sorensen, A., Andrews, L., & Drennan, J. (2017). Using social media posts as resources for engaging in value co-creation. *Journal of Service Theory and Practice*, *27*(4), 898–922. https://doi.org/10.1108/JSTP-04-2016-0080

Steiner, C., & Reisinger, Y. (2006). Understanding existential authenticity. *Annals of Tourism Research, 33*(2), 299–318. https://doi.org/10.1016/j.annals.2005.08.002

Suntikul, W., & Jachna, T. (2016). The co-creation/place attachment nexus. *Tourism Management, 52,* 276–286. https://doi.org/10.1016/j.tourman.2015.06.026

Tashakkori, A., & Teddlie, C. (2016). *SAGE handbook of mixed methods in social & behavioral research.*

Tiberghien, G. (2019). Managing the planning and development of authentic Eco-cultural tourism in Kazakhstan. *Tourism Planning & Development, 16*(5), 494–513. https://doi.org/10.1080/21568316.2018.1501733

TOF. (2020). *Oman.* [online] Timesofoman.com. https://timesofoman.com/article/901134/oman/

Tremblay, P. (2006). *Desert tourism Scoping study. Report 12. Alice Springs: Desert knowledge cooperative research Centre.* http://www.nintione.com.au/resource/DKCRCReport-12-Desert-Tourism-Scoping-Study.pdf

Vada, S., Prentice, C., & Hsiao, A. (2019). The influence of tourism experience and well-being on place attachment. *Journal of Retailing and Consumer Services, 47,* 322–330. https://doi.org/10.1016/j.jretconser.2018.12.007

Vaisnore, A., & Petraite, M. (2011). Customer Involvement into open innovation processes: A conceptual model. *Social Sciences, 73*(3). https://doi.org/10.5755/j01.ss.73.3.793

Vargo, S., & Lusch, R. (2012). *Toward a better understanding of the role of value in markets and marketing.* Emerald Group Pub.

Vargo, S., & Lusch, R. (2014). Inversions of service-dominant logic. *Marketing Theory, 14*(3), 239–248. https://doi.org/10.1177/1470593114534339

Wang, S. (2010). *In search of authenticity in historic cities in transformation: The case of Pingyao.* SSRN Electronic Journal.

WEF. (2018). [online] Ev.am. http://ev.am/sites/default/files/WEF_TTCR_ HYPERLINK "http://ev.am/sites/default/files/WEF_TTCR_2017.pdf"2017 HYPERLINK "http://ev.am/sites/default/files/WEF_TTCR_2017.pdf."pdf

Woyo, E., & Amadhila, E. (2018). Desert tourists experiences in Namibia: A netnographic approach. *African Journal of Hospitality, Tourism and Leisure, 7*(3), 1–13.

Yang, S., Lin, S., Carlson, J., & Ross, W. (2016). Brand engagement on social media: Will firms' social media efforts influence search engine advertising effectiveness? *Journal of Marketing Management, 32*(5-6), 526–557. https://doi.org/10.1080/0267257X.2016.1143863

Index

Note: Page numbers in *italics* refer to figures, those in **bold** refer to tables and endnotes are indicated by the page number followed by 'n' and the endnote number e.g., 20n1 refers to endnote 1 on page 20.

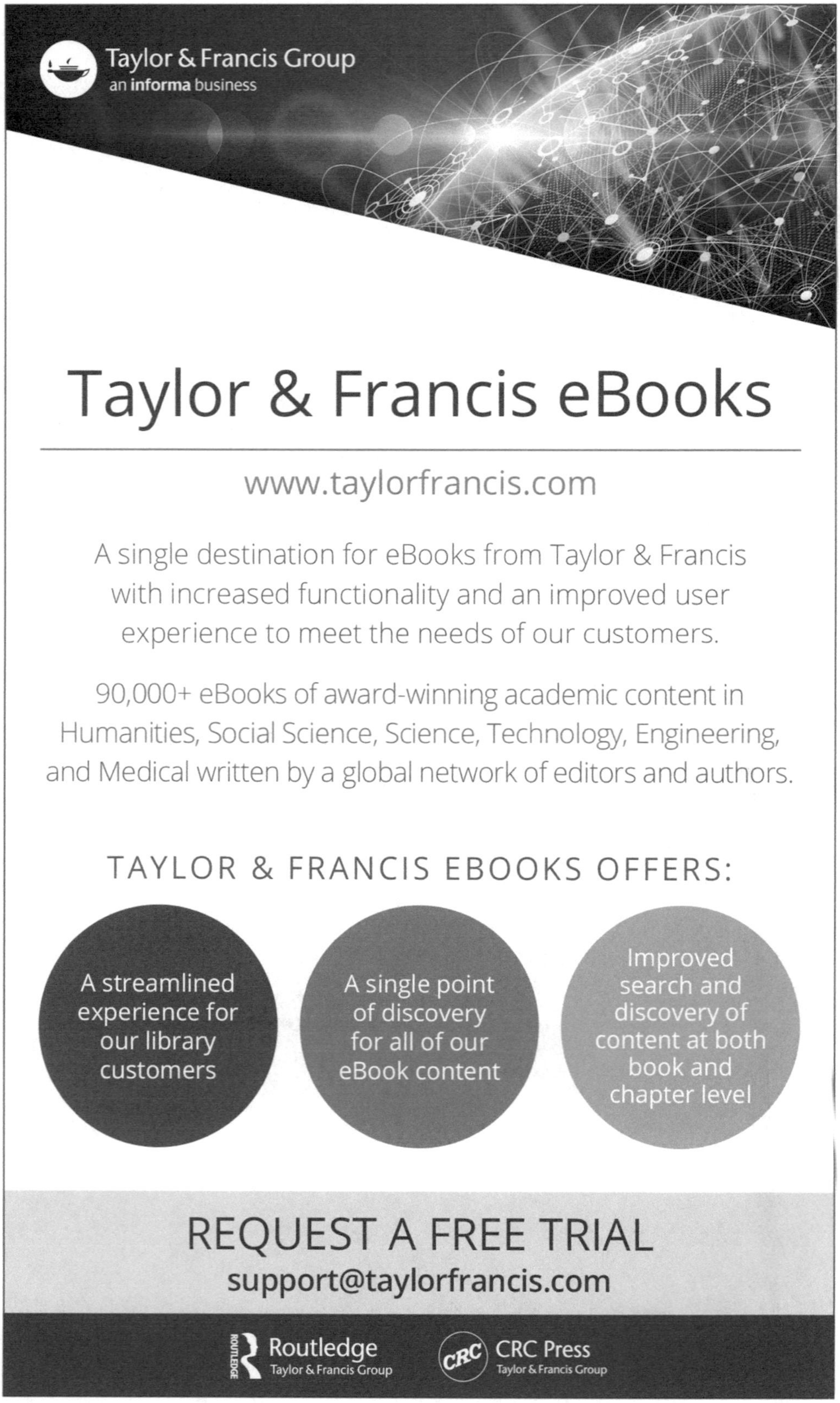